Value Stream Management

Eight Steps to
Planning,
Mapping,
and Sustaining
Lean Improvements

Don Tapping, Tom Luyster,
and Tom Shuker

PRODUCTIVITY
productivity press

Productivity Press • New York

Most Productivity Press books are available at quantity discounts when purchased in bulk. For more information contact our Customer Service Department (800-394-6868). Address all other inquiries to:

Productivity Press
444 Park Avenue South, Suite 604
New York, NY 10016
United States of America
Telephone: 212-686-5900
Telefax: 212-686-5411
E-mail: info@productivityinc.com

Page composition by William H. Brunson, Typography Services
Printed and bound by Malloy Lithographic in the United States of America
Published in the USA by Productivity Press, New York, New York

ISBN: 1-56327-245-8

06 05 04 03   5 4 3 2

Contents

Overview: Value Stream Management

There is a difference between *doing* lean and *being* lean. What exactly does it mean to be lean? In 1990, James Womack and Daniel Roos coined the term *lean production* in their book *The Machine that Changed the World*. Lean production (also known as "lean manufacturing" or just "lean") refers to a manufacturing paradigm based on the fundamental goal of the Toyota Production System—continuously minimizing waste to maximize flow. To become lean requires you to change your mind-set. You must learn to view waste through "fresh eyes," continuously increasing your awareness of what actually constitutes waste and working to eliminate it.

Many manufacturing organizations recognize the importance of becoming lean. However, many organizations are *doing* lean without necessarily *becoming* lean. Typically, such organizations sporadically implement improvements without linking their efforts to an overarching strategy. This book shows you how to make this link by using proven tools. The more you use these tools in your efforts to become lean, the more likely you are to make sustainable improvements.

Value Stream Management Toolkit

QUESTIONABLE, UNSUSTAINABLE IMPROVEMENTS
- Kaizen Workshops (early 1990s)
- Value Stream Workshops (late 1990s)

SUSTAINABLE, REPORTABLE, LEAN IMPROVEMENTS:
- Structured Steps
- Management Commitment Checklist
- Total Employee Involvement
- Part-Routing Analysis
- Value Stream Charter
- Kaizen Meeting Forms
- Demand, Flow, and Leveling Focus
- Route Collection Checklist
- Value Stream Mapping
- Lean Assessment
- Lean Guidelines
- Storyboard
- Kaizen Milestone Worksheet
- Value Stream Sunset Report

What many manufacturers failed to understand in their initial excitement and eagerness to get started with the Toyota Production System is that implementing it involves more than just applying individual methods and tools such as value stream mapping, cell design, kanban, and calculating takt time. All the appropriate tools must be used in such a way that everyone connected within the value stream can work together to improve overall flow to the customer, with little or no waste.

The Purpose of this Book

This is a how-to workbook that helps you immediately integrate the tools shown above into your lean manufacturing improvement efforts. It provides the means for managing the entire lean transformation process effectively to create the kind of enterprise in which continuous learning and continuous improvement go hand-in-hand.

The purpose of this book is twofold:

- To simplify the fundamental lean concepts of demand, flow, and leveling, which you will apply in implementing your lean plan.
- To demonstrate the overall lean *process* that will allow you to accelerate, coordinate, and most importantly, sustain your efforts and assure that everyone is on the same page. This eight-step process is called Value Stream Management.

The Origins of Value Stream Management

Value Stream Management is a process for planning and linking lean initiatives through systematic data capture and analysis. Value Stream Management is a synthesis of best practices used at Fortune 500 companies that have not only successfully implemented lean manufacturing practices, but also sustained them. Divisions within companies such as Daimler-Chrysler, Eaton Aerospace, Delphi Automotive, NUMMI, Thedford Corporation, and Wiremold, Inc., for example, all do an excellent job planning, managing, implementing, and sustaining lean manufacturing improvements. Applying the Value Stream Management process as instructed in this book makes it possible for *any* company to significantly improve its manufacturing processes!

We must, of course, give great credit to Toyota Motor Company for its willingness to share its world-class manufacturing improvement strategies. Through absorbing lessons learned by the companies mentioned above, as well as others, and with support from the Toyota Supplier Support Center, many manufacturers have not only learned more about Toyota practices, but also have adapted the system to meet some of the unique demands present within their own environments.

Who Should Read this Book?

This book has been written for people with various degrees of lean understanding and experience. If you are just starting to learn about lean, this book will teach you an approach that will enhance your future efforts. If you have attended some workshops or simulations, read some literature, or participated in lean improvement teams, this book can change the way you approach becoming lean.

This book is purposefully written for two groups of people with specific needs:

- *Top management* must understand the Value Stream Management process and believe in it before attempting to apply it within the organization. Value Stream Management provides this group with the necessary structure for this commitment as well as a communication tool—the Value Stream Management storyboard—that will satisfy their need for effective metrics and reporting.

- *Supervisors, managers, and team leaders* must understand how to use Value Stream Management for planning and reporting, and also apply it in a way that makes it easier to get products out the door.

This dual perspective allows each group to gain additional insight into the other, and together to leverage the power of Value Stream Management.

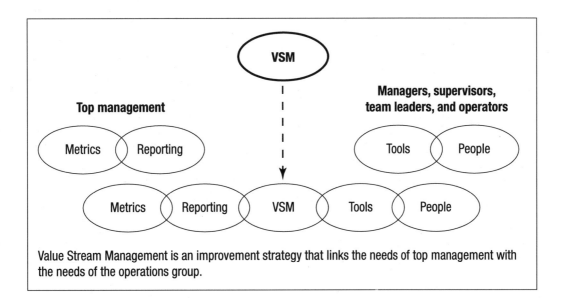

Value Stream Management is an improvement strategy that links the needs of top management with the needs of the operations group.

Learning Features

As you implement Value Stream Management, it's important to ensure that everyone has a good understanding of lean concepts. To help you with this we have included the following learning features:

→ **Guidelines and Checklists**—Lists of questions and guidelines help you plan your lean future state and direct your implementation efforts.

→ **Case Study**—Learning to apply lean manufacturing methods and tools is an adventure. You can and should study the tools and concepts, but the only way you will truly learn them is by applying them. We have tried to *show* you how to apply the major tools and concepts by breaking each of the eight steps of Value Stream Management into smaller steps and including a case study.

→ **CD-ROM**—Over two dozen helpful forms and supplementary worksheets are included on the companion CD-ROM. The "CD" icon is used throughout the text to indicate when you should refer to the CD-ROM for additional forms.

→ **Lean Manufacturing Assessment**—The CD-ROM includes a lean manufacturing assessment that helps you create a snapshot of your organization's current state and understand in a broad sense where you are now and what you need to accomplish to become lean. Performing the lean manufacturing assessment will help you isolate improvement opportunities and suggest metrics to help drive the lean transformation.

→ **Glossary**—In the glossary you will find definitions of common lean terms and concepts used throughout the book.

→ **Bibliography**—The bibliography includes some excellent references on the basic concepts and tools.

The contents of this book have also been adapted into a video training program entitled *Value Stream Management: Eight Steps to Planning, Mapping, and Sustaining Lean Improvements*, which has been thoughtfully designed to guide your core implementation team through the Value Stream Management process by engaging people in more active learning.

We believe that you will find this book an indispensable tool for study and reference as you progress through your transformation to lean.

Introduction: The Value Stream Management Process

- **Where should we start? Where are we in our lean implementation?**
- **Why do we seem to be no further ahead this year than last year?**
- **We have created value stream maps—why haven't we done much with them?**
- **Whatever happened to that lean team that was started six months ago?**
- **We have been doing kaizen blitzes, so why aren't we seeing any big improvements?**

You will find answers to these questions and many more as you go through the detailed analysis and application of the Value Stream Management process. Many organizations have implemented pieces of the process, but few have taken the initiative or spent the time to complete it in its entirety. As a result, relatively few organizations have created a sustainable lean manufacturing system.

Value Stream Management is a process for planning and linking lean initiatives through systematic data capture and analysis. Value Stream Management consists of eight steps:

1. Commit to Lean

2. Choose the Value Stream

3. Learn about Lean

4. Map the Current State

5. Determine Lean Metrics

6. Map the Future State

7. Create Kaizen Plans

8. Implement Kaizen Plans

Value Stream Management is not just a *management* tool; it is a proven process for planning the improvements that will allow your company to become lean.

The Value Stream Management Storyboard

The Value Stream Management process employs the *storyboard*—a powerful tool. It is a poster-sized framework for holding all the key information for planning lean implementation; a small, blank version of it appears on the inside front cover, with a completed version (from the case study) on the inside back cover. Your team will enter data and value stream maps on the storyboard to build a shared document of what you've done and plan to do.

For demonstration purposes, we will show the information being added to the storyboard during each of the eight steps. In practice, many teams find it easier to complete the storyboard during Step 7 to use as a management review document. Be flexible and use the storyboard in a manner that works most effectively for your team.

The storyboard tool is an important reason why Value Stream Management is such an effective path to lean. In a truly lean enterprise, both material and information flow freely. Visual management ensures that everyone knows the organization's goals and that all information needed by people to work as effectively as possible is easily accessible. Storyboards are commonly used at Toyota precisely for this reason—they help people see and understand the "big picture" and buy into the overall strategy.

Although we recommend that you use the storyboard tool, we have also included an alternative communication format for reporting and planning your improvements, which is commonly used in many organizations. It is a "package" of documents that includes a team charter, meeting information form, status reports, and a "sunset report" related to the value stream improvement process.

Why Use Value Stream Management?

Value stream mapping has become the latest craze in manufacturing improvement. This is exciting, because value stream maps are an important part of what makes the storyboard an exceptional form of visual management. However, it's not enough to perform mapping in isolation. Without a good understanding of lean manufacturing principles, mapping will bring organizations no closer to minimizing wastes and achieving excellence than kaizen workshops did in the early- to mid-nineties. While manufacturers have been quick to acknowledge the benefits of becoming lean, relatively few truly understand what such an effort entails.

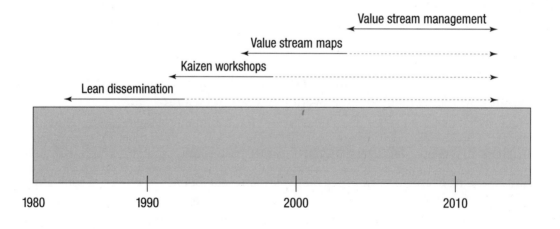

Critical lean elements often are missing in current applications. If you hope to create an authentic lean enterprise, rather than a superficial one, you must learn the tools and methods of lean and how to integrate them. What is needed is a complete process that links strategic plans to daily work while at the same time teaching the fundamentals. The eight steps of Value Stream Management, followed sequentially, is that process. In addition, we have learned through experience that a successful lean manufacturing initiative depends on four critical behaviors:

❏ Make a true commitment.

❏ Understand customer demand thoroughly.

❏ Depict the current state accurately.

❏ Communicate, communicate, communicate!

Make a True Commitment

Is achieving a lean enterprise important enough to allocate the resources for effective planning, implementation, and maintenance? Your answer should be a resounding yes! Lean results cannot be realized without commitment from people at all levels of the company.

It's not enough to appoint one person to be responsible for lean initiatives and to conduct kaizen workshops or generate value stream maps. You must generate a desire to improve that fuels all other activities. The more people believe and do, the more people *will* believe and do. The guidance and commitment must be sincere and must start at the top. No single tool, methodology, or process will achieve this for you. It must come from a burning desire *to be lean!* Making and sustaining the commitment to the lean transformation is important throughout the Value Stream Management process, of course, but is emphasized in Step 1 and in Step 8.

Understand Customer Demand Thoroughly

Variation in customer demand is not a reason for avoiding true lean implementation—only an excuse. It may take a little more work than expected, but you *can* understand customer demand and incorporate it into your lean process. Analyzing customer demand is a particularly important concern when you are choosing the target value stream (Step 2), mapping the current state (Step 4), and mapping the future state (Step 6).

Common Excuses for Not Implementing Lean

"Our customer requirements are too complicated to utilize the Toyota Production System—that works only for automotive."

"We have too many unique requirements throughout our processes to use lean."

"None of our orders are the same."

"We have over 6,000 variations in our product line!"

Depict the Current State Accurately

Before you begin implementing lean, you must fully understand what you are currently doing in relation to cycle times, process communications, people's work standards, machine/equipment capacity, and so on. Only by grasping the present conditions can you create a future condition and plan how to implement it.

A current-state map is a snapshot of conditions at a specific point in time. Making such a map may require a full day of team involvement—and remember, this is just to see the current state. Do not underestimate this important process! Even though you may want to jump to creating the future state, be careful about the assumptions you may be making. If you haven't depicted the current condition accurately, you will have significant problems later in implementation. Be accurate and precise! Do not rush when collecting this information.

Communicate, Communicate, Communicate!

Manufacturing professionals talk about and understand the importance of driving fear from the workplace and creating a "no-blame" environment. However, in practice, most organizations are far from perfect in this area. This is why it is extra important to make the effort to treat everyone with dignity and respect. Good communication is essential to this effort. Telling people what you are doing and why—and expressing a sincere interest in making sure they understand—does much to create an environment befitting a lean enterprise. The more you communicate with people in this manner, the more you will earn their trust and gain their enthusiastic support.

We have already talked about the importance of visual management and visual communication in implementing lean. Remember that good face-to-face communication is equally important. It establishes the rapport that makes people receptive to actually using the visual tools.

Attributes of Value Stream Management

The Value Stream Management process supports the transformation into a lean enterprise by providing a structure to ensure that the lean implementation team functions effectively. That structure, made visual through the storyboard format, encompasses the strengths of proven problem-solving methods:

- ✓ It provides for clear and concise communications between management and shopfloor teams about lean expectations and about the actual material and information flow.
- ✓ Proven tools are used for implementation.
- ✓ Team recognition and ownership are included from beginning to end.
- ✓ Management review and reporting are incorporated.
- ✓ It provides a good form of visual communication
- ✓ Changes and updates can be reflected as they occur.

What Value Stream Management Is—and Is Not

Any proven process can fail to achieve results if people do not apply it properly or if they lack a fundamental understanding of the nature of the process. Here are some key points to remember about Value Stream Management.

Value Stream Management is a process that:

✓ Links together people, lean tools, metrics, and reporting requirements to achieve a lean enterprise.

✓ Ensures that lean is sustained.

✓ Allows everyone to understand and continuously improve his or her understanding of lean concepts.

✓ Makes possible controlled process flow on the floor.

✓ Generates an actual lean design and implementation plan.

✓ Requires a lean coordinator to make the process go smoothly.

Value Stream Management does not involve:

✗ Just forming kaizen teams and waiting for results.

✗ Just mapping the value stream to show material and information flow.

✗ Just forming self-directed work teams and waiting for results.

✗ Appointing improvement coordinators or "lean coordinators" and making them responsible for improvements.

Most important, Value Stream Management is *not* a method for *telling* people how to do their jobs more effectively. It is systematic approach that empowers people to *plan how and when they will implement* the improvements that make it easier to meet customer demand. Value Stream Management is not about making people work faster or harder; it is about putting in place a system so that *material* flows through manufacturing processes at the pace of customer demand. This book shows you how to make this happen.

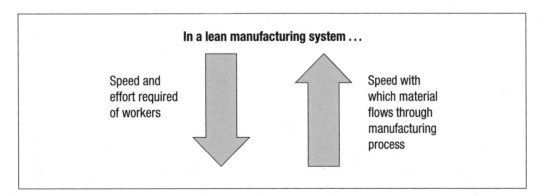

Value Stream Management encompasses all the functional and operational relationships that exist within the value stream. It addresses the business impact of a lean transformation, promotes union-management cooperation, and clarifies what lean means to the different constituents.

Put People First

We have found that simply applying tools such as value stream mapping, supermarkets, heijunka, u-shaped cells, and point kaizen workshops *in isolation* does not necessarily produce significant *sustained* changes in the flow. But organizations that allow people to be part of the kaizen culture—part of the design of the future state, part of the overall planning process—will certainly come closer to attaining world-class status than those that do not. Value Stream Management helps ensure that people are considered first and foremost in lean activities.

People, and their efforts to eliminate waste, are critical to successfully implementing and sustaining a lean system. If the well-being of its personnel is not a priority for the organization, then a true kaizen culture is unlikely to evolve. Value Stream Management will greatly assist in this, but in and of itself it does not provide all the answers. Value Stream Management only works as well as management's ability to truly foster people's well-being. Throughout each step we will focus not only on the physical aspects of lean implementation, but on the human issues as well.

Key Lean Principles

As you go through the eight steps of Value Stream Management, keep in mind the following lean management principles:

❑ Define value from your *customer's* perspective.

❑ Identify the *value stream.*

❑ Eliminate the seven deadly *wastes.*

❑ Make the work *flow.*

❑ *Pull* the work, don't push it.

❑ Pursue to *perfection.*

1. Commit to Lean

Step 1. Commit to Lean

To many people, lean implementation will look like just another program. It's up to management to lead the commitment to lean and to demonstrate how and why it is different. The entire organization's commitment to lean will mirror the commitment of top management.

World class is a never-ending journey, not a destination. A world-class organization:

- ❑ Operates by the cost-reduction principle.
- ❑ Produces the highest quality in its business sector—zero defects.
- ❑ Meets quality, cost, and delivery requirements.
- ❑ Eliminates all waste from the *customer's* value stream.

Toyota describes this direction as "True North." You must stay the course. Lean is not a compromise: it is the continual effort throughout the journey, embracing tools, support systems, and *people* that allows organizations to achieve world-class status.

Management Push or Worker Pull?

In creating change, we often talk about the difference between a *management push* system and a *worker pull* system. Management push—the traditional change strategy in many companies—means management's ordering or "pushing" improvement activities on employees. Worker pull, on the other hand, means motivated employees pulling the resources and training they need to improve the value stream. This is the preferred method in companies aspiring to become lean.

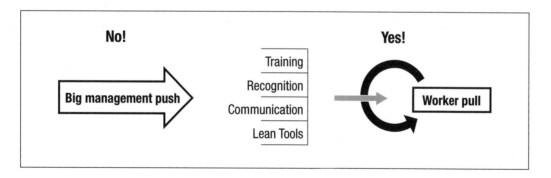

In a worker pull system, improvements and cost-reduction ideas come "naturally" from the people who are most familiar with the processes. However, a natural worker pull system will not develop without management's direction, guidance, and support. "Management" includes supervisors, planners, process engineers, lean implementers, and team leaders. The decisions these people make every day affect the flow of information and material. They are the "backbone" for lean implementation—the structure that supports the "living" value stream. The people on the shop floor are the "organs" for lean, performing the vital functions that sustain and improve the system.

Many lean initiatives fail early on because management has not committed to the project but assigns a team anyway. The Value Stream Management process ensures a greater chance of lean success by generating commitment from management throughout the eight steps.

Catchball

Lean companies differ most radically from traditionally run manufacturing companies in that information flows freely in many directions. In fact, information flow is instrumental to management's demonstrating an authentic commitment to becoming lean. The beauty of this is that commitment grows stronger when it flows both from the top down and from the bottom up. The process of *catchball* makes this possible.

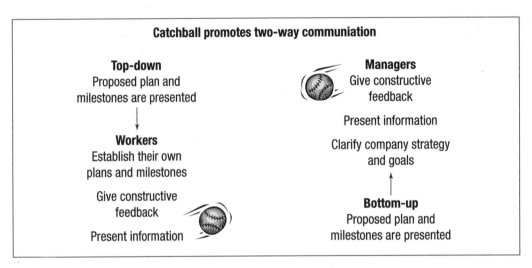

Catchball is simple. Regardless of who initiates a project, that person articulates the purpose, objectives, and other ideas and concerns, and then "throws" them to other stakeholders. In Value Stream Management, the catchball process begins as soon as a manager assembles a core implementation team and identifies an area to improve. Based on the purpose, objectives, and concerns communicated by the manager, the team completes a team charter that defines the project in more detail—then throws it back to the manager. The catchball process continues until management approves the charter. Catchball is also used to come to agreement on metrics (Step 5), the future-state map (Step 6) and kaizen plans (Step 7).

Catchball accomplishes three things:

✓ It ensures that management is committed to the core implementation team's ideas.

✓ It ensures that everyone who should give input does so.

✓ Most importantly, it establishes a credible and reliable structure for workers to initiate improvement.

Catchball is the heart of a worker-pull system of improvement.

Key Management Activities

To reduce and eliminate waste effectively, employees must be behind the lean transformation effort 100 percent. Ensuring their support starts with communication between top management and all levels of the organization. Before beginning actual activities, as well as during these activities, top management must articulate the need to become lean. Management can do this by:

❑ Holding monthly, bimonthly, or quarterly meetings to inform people about new customer requirements, increases in raw material prices, new capital expenditure requirements, and so on.

❑ Demonstrating the pricing of current competitive products and sharing margins when possible.

❑ Displaying customer letters, both positive and negative, to all manufacturing and non-manufacturing areas.

❑ Establishing administrative lean teams to acknowledge that improvement is everyone's concern, not just production's.

Once everyone understands the need, top management must find ways to open doors, allowing others in the organization to contribute to their full capacity. This is accomplished through four main steps, each of which we will now detail.

1. Identify the Value Stream Manager/Champion and Core Implementation Team Members

Top management's first task is to select a value stream champion (or manager) and make sure this person clearly understands the need for lean transformation. He or she is the expression of management's commitment. The value stream champion (or manager) should have the authority and responsibility to allocate the organization's resources. Likely candidates include the manufacturing manager, product manager, or process manager; in a smaller organization, it may be the plant manager or general manager. Rarely is the champion an internal lean manufacturing expert. He or she reports directly to top management, usually to the plant manager or general manager.

A value stream champion should possess the following attributes:

❑ A sense of product ownership.

❑ A sense of commitment to lean manufacturing.

❑ Authority to make change happen across functions and departments.

❑ Authority to commit resources.

Once the champion has bought into the project, he or she will:

- ❏ Help select core implementation team members and introduce them to the Value Stream Management process.
- ❏ Monitor the team's progress in applying the eight steps.
- ❏ Be available to remove roadblocks as they arise. This does not mean attending every meeting, but it does entail going to the floor often to see what the team is working on and listening to their ideas, plans, and needs.
- ❏ Review the future-state map and kaizen proposals.

In addition to identifying the champion, it's important to assemble a core implementation team for Value Stream Management. The core team will take ownership of the process and reach consensus on a team charter, following the direction of the champion. The team will:

- ❏ Create plans and maps.
- ❏ Communicate to all levels within the organization.
- ❏ Make sure people are trained.
- ❏ Implement the Value Stream Management process.

It's critical for team members to work well together, because every aspect of Value Stream Management requires a high degree of collaboration—especially mapping the current state and designing the future state.

Top management often assigns people to the core implementation team. The team should consist of three to seven members. The team is cross-functional; its members should be a good representation of the people who will be expected to sustain the system. Team members may include process engineers, continuous improvement facilitators (or "lean representatives"), team leaders, planners, human resources specialists, engineers, accounting/financial specialists, experienced process operators, and internal lean manufacturing specialists.

The Cross-Functional Team

Why would you include someone from HR or Finance on the core implementation team? Remember: as you go lean, the flexibility of the workforce will be critical. Your lean transformation may require redefining or reclassifying jobs, or even deploying people to different areas. This goes more smoothly when HR understands the reasoning behind the changes. What's more, you will need the support of your financial people for the metrics you decide to use. Throughout the lean journey you may need to request funds to implement major initiatives. Having financial people on the team helps them understand the results that lean initiatives can and will achieve.

While you do not need to include someone from every process, it is important to have on the team individuals who can communicate with every area, including administrative areas.

The core team should follow basic teaming guidelines:

❑ Identify team roles, such as leader, scribe, timekeeper, and facilitator.

❑ Establish team norms.

❑ Include the value stream champion in the first meeting.

The team leader performs several key functions at meetings and in between:

❑ Supports the team members—the people—throughout the process.

❑ Schedules meetings.

❑ Uses the storyboard tool (or appropriate forms) to communicate the team's mission and progress to all value stream parties.

❑ Brings people with additional expertise into the team process as needed.

❑ Communicates with the champion and plant manager on a regular basis.

❑ Understands team dynamics and the teaming stages and is alert for signs of resistance, which can be caused by inadequate knowledge about lean manufacturing principles or by not clearly articulating the need for the lean transformation.

❑ Addresses nonparticipation early and privately.

2. Kick Off the Value Stream Management Project

It is important to kick off the Value Stream Management project by giving the team as much clarification as possible at the first meeting. The champion should attend the core team's first meeting to explain:

❑ Why the team was assembled.

❑ How team members were selected.

❑ The need for applying lean principles and tools.

❑ The reasons for choosing the area of focus (i.e., competitive products on the market, customer demand for price reductions, customer demand for lead-time reductions, significant quality problems attributed to manufacturing processes, etc.).

❑ How the team and the project will support corporate strategy and goals.

Make sure team members understand their overall objectives and the initial area of focus, and know that they can add and remove members as needed to reach those objectives.

Other areas to address at the kickoff meeting include:

❑ A review of the Value Stream Management process and what it means to the team.

❑ Expected duration of the project.

❑ Expected communications.

❑ Resource allocations to accomplish the objectives.

❑ Questions from team members.

The core implementation team should document its formation and the scope of its work with a charter. A team charter is a document that lists the team members and their roles, and articulates the team's purpose, the expected duration of the project, the resources available, the range and boundaries of the project, and the mechanism for reviewing the team's work. The team charter is a "living document" that should be updated to reflect changes as they occur (for example, the addition of team members, a change in the scope of the projects or metrics, etc.).

As the team meets and begins working toward its goals, it should document its efforts. The worksheets on the accompanying CD-ROM provide good examples of documents that encourage structured meetings and actionable outcomes.

3. Go to the Floor!

Most organizations build invisible walls that separate functions or departments, making communication and teamwork difficult. Even in a completely "open" environment, with all functions under one roof, walls may still exist.

Breaking down these barriers is one of top management's most important jobs. Before this can happen, though, managers must understand their organization's activities by "going to the floor" and observing production first-hand. It is more effective for management to go to the "parts world" on the manufacturing floor than for machine operators and assemblers to come to the "office world." Remember that those operators are busy creating value for customers—getting parts out the door on a production schedule. Management, supervisors, and team leaders must go to the floor and aggressively address employee concerns.

Other staff members also need to increase their awareness of the issues shopfloor workers face. Many lean organizations physically relocate planners, quality technicians, process engineers, and lean facilitators and trainers on or near the shop floor. This allows for better communication and rapid feedback when problems or concerns that disrupt product flow arise.

A Question for Top Managers

Where do you park your car? Consider parking where the production employees park. In addition to reducing some of the status barriers that block communication, this will require that you spend some time each day walking through the factory to reach your office. This simple change can have a profound positive impact.

---

4. Review All Value Stream Kaizen Proposals

After analyzing the current state and mapping a future state, the team presents kaizen proposals to management in Step 7. Management must review their ideas with great respect and care. The team has spent considerable time and effort to get here, and has overcome many obstacles. Key elements to remember during your review:

❑ Thank the team for its hard work.

❑ Gain an understanding of the team's plan and rationale.

❑ Ensure consensus has been achieved and inquire whether everyone connected to this value stream is aware of the proposal.

❑ Offer additional resources to assist the team, if needed. This may involve outside consulting services, benchmarking trips to other plants, and so on.

❑ Thank the team again for its hard work.

Reaffirm for team members how their work is supporting corporate-level strategies and goals. Explain how their improvements are helping to strengthen the entire company—not working people out of jobs. Finally, focus on what you can do to make their jobs easier in improving the value stream.

Invest in Your People

Your company's manufacturing or assembly workers are the experts in your plant. They are familiar with every hidden detail of their work and how to improve it. Management must organize a structure, encourage communication, and support their ideas for improvement. This is what the Value Stream Management process and storyboard promote.

To progress toward lean, management must treat employees as human "fixed assets." A fixed asset appreciates in value over time. Unlike equipment assets that lose value every day, the experience, training, and technical expertise of your employees increases in value (Figure 1-1).

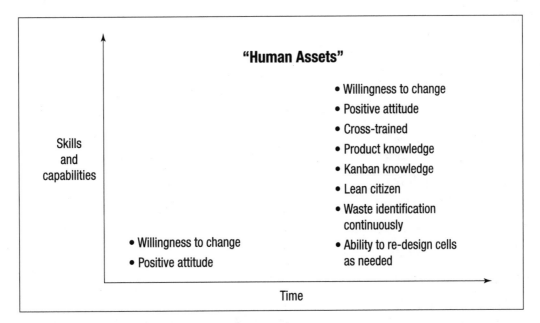

Figure 1-1. Human assets appreciate in value over time

Invest in your people and make them secure about their activities. As they make improvements and take costs out of the value stream, they need to feel that their jobs are not at risk. Practice mutual trust and respect every day. The more you support people through *word and deed*, the more buy-in and support they will contribute toward cost reduction and waste elimination.

Keeping focused on manufacturing processes is an important part of overall efforts to improve a value stream, but you eventually must also focus on *administrative* processes to meet overall customer cost, quality, and delivery requirements. Figure 1-2 shows an inclusive view of all processes within a typical value stream that must be improved in order to reduce lead times and meet customer requirements.

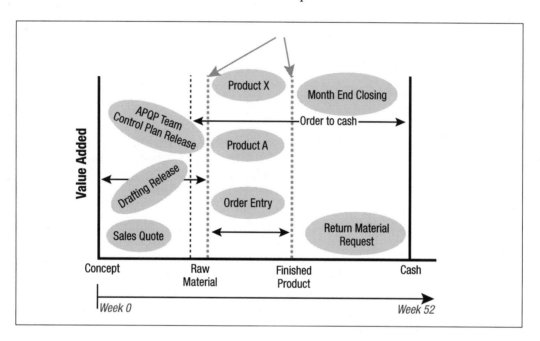

Figure 1-2. Typical manufacturing value stream processes

Short-Term Pains and Long-Term Gains

Management must recognize that the transformation to lean may involve some short-term pains, which could test their level of commitment. Short-term pains may include:

X Additional tooling/fixtures required to reduce setup times.

X Overtime to rebalance lines and educate the workforce.

X Smaller trays/carts for small-lot production.

X Initial buffer and safety inventory.

X Engaging a lean expert to accelerate your efforts.

X Benchmarking to "see" (or understand) how others have achieved success.

X Managers/supervisors/team leaders spending time to thoroughly understand lean.

X Reorganization of the facility.

However, you can expect to see the following long-term gains:

- ✓ Work-in-process: from weeks to days.
- ✓ Defect rate: from 3 sigma to 6 sigma (from 10,000 defective ppm to 3.4 defective ppm).
- ✓ Value-added ratio: 500 percent increase.
- ✓ Changeover reduction: from hours to minutes.
- ✓ Overall equipment effectiveness: 40 percent increase.
- ✓ Process routing: from greater than 1,000 feet to less than 20 feet.
- ✓ Factory floor space: greater than 50 percent reduction.
- ✓ People: significantly increased expression of latent creative potential.

There are other tangible benefits of becoming lean:

- ✓ Makes work safer and easier.
- ✓ Sharpens perception.
- ✓ Promotes cooperation.
- ✓ Shortens feedback loops.
- ✓ Speeds corrective action.
- ✓ Speeds learning.
- ✓ Improves process reliability.

There is no doubt that the gains associated with becoming lean are worth it. All types of industries can benefit from a lean transformation, including aerospace, automotive, chemicals, industrial, pharmaceutical, machine tools, consumer products, and job shops.

Implementing Lean Transforms a Business Culture

An organization can fear its competitors—or it can become fiercely competitive by putting energy confidently and quietly into aggressive lean implementation. You can transform fear into a positive impulse by viewing competitive challenges as an opportunities to change the way business is being conducted, and by working to implement change in a controlled way. Applying Value Stream Management greatly enhances your ability to accelerate lean implementation and to control change rather than letting change control you.

Many organizations attempt to create a competitive attitude by trying to change the culture itself. They assign everyone to teams, "empower" people, implement self-directed work teams, appoint team leaders, and so on. It is difficult to build a competitive culture that way.

Value Stream Management takes a different approach. It is structured to support a *change in behavior* by applying lean concepts and tools. Then, through proven success with results, *employee attitudes will start to change*. And when those pockets start to get larger throughout the organization, your culture will have shifted to lean. This is one of the reasons that Toyota is now a revered, world-class manufacturing company.

Success comes through learning from failures—which are more appropriately viewed as "experiments in improvement." Improvement ideas need to be tested and controlled experimentally, without disruption to adjoining processes. Not every lean tool will work on the first try. Lean success is an iterative process: two steps forward, one step back, and two steps forward again. This is one of the key elements within the Toyota Production System: continual improvements, constantly tested and adopted, causing no disruption to the customer.

Commitment Checklist

Lean implementation is a simple concept that is challenging to implement and even more challenging to sustain. It demands disciplined and passionate people to lead the charge. Transformation to a lean state does not occur randomly or naturally in organizations; it requires diligent management commitment and detailed planning.

Is Management Committed to Lean?

Management shows its commitment to lean by:

- ❏ Establishing and maintaining clarity of purpose.
- ❏ Committing human resources to lead daily lean activities.
- ❏ Allocating the time and resources for training.
- ❏ Assuring that the team possesses working knowledge of lean tools.
- ❏ Constantly communicating with the team and monitoring its activities.
- ❏ Removing the roadblocks that hinder the team's progress.
- ❏ Allocating appropriate dollars within a short time period.
- ❏ Providing clear incentives for the team's success.
- ❏ Staying flexible with the dates and times of the project.
- ❏ Ensuring total employee involvement.
- ❏ Participating actively throughout the Value Stream Management process.

You know you *do not* have a commitment when management:

- ✗ Repeatedly postpones the kickoff meeting.
- ✗ Does not attend, or does not communicate at, the team's kickoff meeting.
- ✗ Does not allocate time for training or benchmarking.
- ✗ Expects that this is just another process improvement team to meet some corporate goal.
- ✗ Provides no additional rewards or incentives.
- ✗ Does not respond to requests for tooling, fixtures, or equipment within an expected time frame (e.g. 30 days). (Here we are not talking about large purchases such as CNC machines.)
- ✗ Shows little interest in what the team is doing, or communicates with the team infrequently.

Throughout the eight steps of the Value Stream Management process, we will refer to a case study involving Premiere Manufacturing, Inc., to demonstrate how the process is implemented. Premiere is an actual company that applied the eight steps; we have changed the name and simplified some facts for clarity.

PREMIERE MANUFACTURING CASE STUDY, STEP 1

Background: The Present Situation

Premiere Manufacturing, Inc., is a Tier 1 automotive supplier producing #4, #6, #8, and #10 coolant hoses for six key customers. Over time, Premiere has been making improvements through kaizen activities. Recently the company replaced two screw machines with two dual-spindle CNC machines, which have not proven to be as reliable as expected.

One of Premiere's key customers, Cord, Inc., has been consolidating its supply base. Cord has been auditing and analyzing their suppliers' abilities to meet quality and delivery demands while also reducing costs. Premiere's long-term agreement (LTA) with Cord is about to expire. Cord is willing to keep doing business with Premiere but is demanding:

- A 5 percent cost reduction annually over the next two years.
- On-time delivery of 98 percent or higher.
- A lead-time reduction from six weeks to less than two weeks.

The plant manager decides to sign the new LTA immediately, because he does not want to run the risk of losing Cord's business. He also has confidence that his people can apply lean tools and methods to achieve the performance goals that Cord demands.

Team Formation and Kickoff Meeting

The plant manager assembles a team that consists of the following members:

- Plant manager: value stream champion.
- Manufacturing manager: team leader.
- Machining supervisor: core team member.
- Planner: core team member.
- Operator cross-trained in several major operations: core team member.
- Product engineer: core team member.
- Internal lean manufacturing specialist: core team member/facilitator.

The plant manager explains to the team that applying lean practices will produce the results necessary to assure a good profit margin. He also lets the team know that he has signed the LTA, so there's no turning back. Premiere has made a commitment to its customer, and now it is the core team's job to make the changes necessary to honor that commitment.

The plant manager also covers the following points during the kickoff meeting:

- He explains the Value Stream Management process and its importance and outlines the time frame for the team's activities.

- He explains how team members were chosen and tells them that each member is critical to successfully meeting the terms of the recently signed LTA.

- He expresses keen interest in reviewing their kaizen proposal once they have completed it.

- He solicits questions and asks the team what he can do as champion to support its efforts.

The plant manager promises to monitor the team's progress and to visit the floor frequently to address team members' concerns. He discusses his expectations regarding how the team will report on its activities and gives an overview of how it will refine the area of focus in Step 2.

The team's first action after the kickoff meeting is to create a team charter that incorporates all the relevant information communicated. The team begins a storyboard, with each team member's name written at the top in the appropriate spaces.

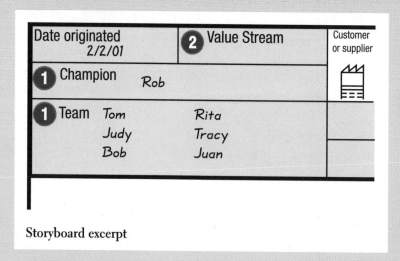

Storyboard excerpt

2. Choose the Value Stream

Step 2. Choose the Value Stream

What Is a Value Stream?

Manufacturing companies survive because they transform raw materials into finished goods that their customers value. Processes transform material into products; operations are the actions (cutting, heating, grinding, bending, etc.) that accomplish those transformations. Operations are considered the process elements that add value, but processes also include non-value-adding elements. A value stream consists of everything—including non-value-adding activities—that makes the transformation possible:

→ Communication all along the supply chain regarding orders and order forecasts. (For example, for a value stream at a first-tier automotive supplier, communication takes place between the first-tier supplier and one or more second-tier suppliers; and between the first-tier supplier and its customer—the original equipment manufacturer.)

→ Material transport and conveyance.

→ Production planning and scheduling.

→ The network of processes and operations through which *material and information flows in time and space* as it is being transformed.

There are many value streams within an organization, just as there are many rivers flowing into an ocean. Value Stream Management helps you systematically identify and eliminate the non-value-adding elements from your value streams.

Selecting Value Streams for Improvement

Your customers often define your value streams. If you make similar parts for a variety of customers (for example, rearview mirrors for Toyota, Ford, and Daimler-Chrysler), each having unique specifications, you would have three product families, each with its own value stream.

If your customer has not defined the value stream for you, there are two reliable methods you can employ to help you decide which value stream(s) to target for improvement:

❏ **Product-quantity (PQ) analysis.** Start with PQ analysis first to see if some part numbers are run in volumes high enough to make the choice an obvious one.

❏ **Product-routing (PR) analysis.** Use product-routing analysis if results from PQ analysis are inconclusive.

Using PQ Analysis

PQ analysis is used to display the product mix as a Pareto chart. A Pareto chart graphically demonstrates the Pareto principle—also known as the 20:80 rule—and helps separate the "critical few" from the "trivial many." The chart shows how the total quantity of products is distributed among different product types, with the assumption that the higher-volume products are the first that should be targeted for improvement.

To perform PQ analysis, follow these steps:

1. Obtain three to six months' worth of data on production output.

2. Enter your products by quantity (from greatest to least) on a PQ analysis list (see Figure 2-1).

PQ Analysis List					
Performed by: D. Arroyo & D. Herrera					Date: 11/16
#	Item (part #)	Quantity	Cumulative Quantity	%	Cumulative* %
1	W	29,000	29,000	41	41
2	R	26,500	55,500	37	79
3	Y	3,000	58,500	4	83
4	I	3,000	61,500	4	87
5	P	2,000	63,500	3	90
6	A	2,000	65,500	3	93
7	D	1,500	67,000	2	95
8	G	1,500	68,500	2	97
9	J	1,000	69,500	1	99
10	L	1,000	71,500	1	100

*Percentages in "Cumulative %" column have been rounded off.

Figure 2-1. PQ analysis list

3. Create a Pareto chart (see Figure 2-2) using the data from the PQ analysis sheet.

4. Analyze the product mix.

In the example shown in Figure 2-2, the combined quantity of two products, W and R, together make up close to 80 percent of the total quantity produced. If the product to quantity ratio is approximately 20:80 (in other words, 20 percent of the product types account for 80 percent of the total quantity of parts produced), you have a high-volume, low-variety product mix on which you should focus value stream improvement efforts. In this example, the targeted value stream would be the process or processes that make products W and R.

If your PQ analysis showed a 40:60 PQ ratio (40 percent or more of the product types account for 60 percent of the total quantity of parts produced), that would indicate a high variety of product types with a relatively low volume of each type (see Figure 2-3). Under these circumstances, your choice is not as obvious, and additional analysis is probably necessary.

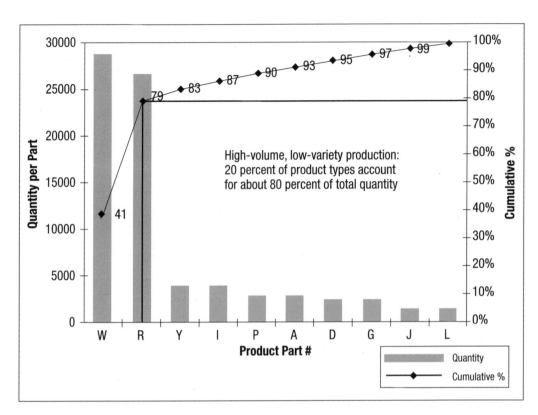

Figure 2-2. Pareto chart—PQ analysis showing a 20:80 ratio

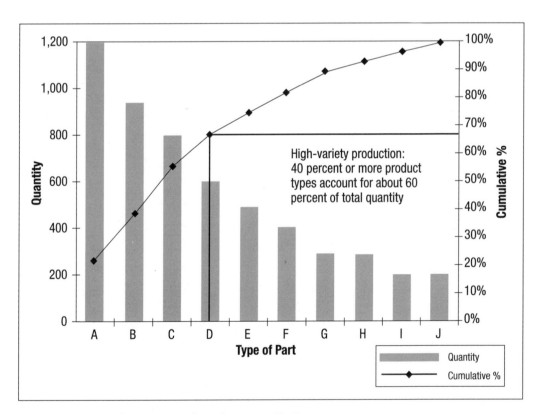

Figure 2-3. Pareto chart—PQ analysis showing a 40:60 ratio

Using Product-Routing Analysis

If your PQ analysis indicates a 40:60 PQ ratio, use product-routing analysis to help you choose your target value stream. In product-routing analysis, you make a chart that shows which products or parts have similar process routes. Products that are processed through the same machines or operations in the same sequence are good candidates for grouping into product families.

To perform PR analysis, follow these steps:

1. Start by showing the process sequence—the sequence of operations—for each product type listed by volume, as shown in Figure 2-4.

Sequenced by Volume

Volume	Product							
20,000	W	rc	c	m	d	od	g	i
12,000	R	rc	c		d	od		i
10,000	Y		c	m	d			i
3,600	I	rc	c	m	d	od	g	i
3,300	P		c		d		g	i
3,100	A	rc	c	m		od		i
2,600	D	rc	c	m	d	od		i
2,300	G	rc	c	m	d	od		i
2,100	J		c	m	d			
1,000	L	rc	c	m	d	od		i

Machine #: 10A   31   32   70B   34   A-10

rc=rough cut • c=cut • m=mill • d=drill • od=outside diameter • g=gauge • i-inspect

Figure 2-4. Sequence of operations by volume

2. Next, group together the products that have the same process routes (Figure 2-5).

3. Analyze the mix of process routes.

As the matrix in Figure 2-5 shows, products *D*, *G*, and *L* have the same process route. Because these products share the same operations, you *might* want to focus on improving a value stream dedicated to producing *D*, *G*, and *L*. However, there are other considerations.

Note that products *W* and *I* have the same process route; so do products *Y* and *J*. Each combination of these products is produced in higher volume than the combined volume of products *D*, *G*, and *L*.

Products with the same process route (i.e. value stream)	Total volume required
W & I	23,600
Y & J	12,100
D, G, & L	5,900

Grouped by Process

Volume	Product							
20,000	W	rc	c	m	d	od	g	i
3,600	I	rc	c	m	d	od	g	i
10,000	Y		c	m	d			i
2,100	J		c	m	d			i
2,600	D	rc	c	m	d	od		i
2,300	G	rc	c	m	d	od		i
1,000	L	rc	c	m	d	od		i
12,000	R	rc	c		d	od		i
3,300	P		c		d		g	i
3,100	A	rd	c	m		od		i

Machine #: 10A   31   32   70B   34   A-10

rc=rough cut • c=cut • m=mill • d=drill • od=outside diameter • g=gauge • i-inspect

Figure 2-5. Sequence of operations by process in identifying value stream

Which product mix represents the value stream that *you* would target for improvement? Most likely you would have to consider several factors. For example, products *D*, *G*, and *L* represent the lowest-volume product mix. But what if these products are manufactured for a customer that has promised you additional business—provided that you can achieve a 5 percent price reduction and 100 percent on-time delivery and zero defects over the next year on the parts you currently supply? Under these circumstances, you might choose to focus improvement efforts on the low-volume value stream among the three being considered.

Additional Considerations for Value Stream Selection

In general, when just getting started with Value Stream Management, you may not need to use PQ analysis or product-routing analysis. Another option is using a simplified version of PQ analysis (Figure 2-6).

The simplified version of PQ analysis is similar to the traditional version in that you list your products along the bottom axis of a bar graph by quantity (from greatest to least). Then you choose a selection indicator. For example, you might decide that any value stream producing a product quantity close to 20 percent of the total volume is a worthy candidate on which to focus improvement efforts. In Figure 2-6, there are two products, W and R, that meet—roughly—the 20 percent criteria. Product W represents 24 percent of total volume; product R represents 19 percent of total volume.

However you decide to choose your target value stream, we recommend selecting one that is neither too simple nor too complex. Of course, the appropriate selection depends on your plant and on customer demand. However, we can offer a few rules of thumb:

❑ Choose a value stream that includes *no more than one machining operation.*

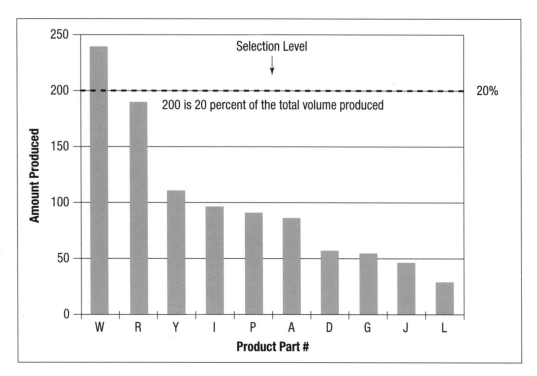

Figure 2-6. Example: PQ analysis simplified version

❏ Choose a value stream that includes *no more than three raw material suppliers*.

❏ Choose a value stream that includes no *more than twelve operations or process stations*.

If you have a high-variety, low-volume product mix, then you will need to use the more sophisticated methods of value stream selection discussed in this chapter.

PREMIERE MANUFACTURING CASE STUDY, STEP 2

The long-term agreement that Premiere has made with Cord commits Premiere to improving on-time delivery and cutting costs associated with the manufacture of #4, #6, #8, and #10 coolant hoses. In effect, this means that Cord has selected the target value stream on which Premiere needs to focus. Nevertheless, the core team performs a product-routing analysis to form a clearer picture of the process routes for each product type.

The product-routing analysis reveals some important facts:

- Products #4 and #6 share the same process route, while products #8 and #10 share a slightly different process route.

- The process route for products #4 and #6 is slightly less complex than the process route for products #8 and #10.

- The total volume of products #4 and #6 is 80 percent of the total quantity produced for Cord.

Volume	Part	Machining	Deburring	Crimping	Welding	Testing	Marking
20,160	#4	●	●	●		●	●
10,080	#6	●	●	●		●	●
4,360	#8	●	●	●	●	●	●
3,200	#10	●	●	●	●	●	●

Premier Manufacturing, product-routing analysis (grouped by process)

In addition, analysis of ordering patterns over the past three months shows that demand for #4 and #6 hoses is nearly constant, whereas demand for #8 and #10 hoses is more sporadic. Given these facts, the team reasons that it is best to focus first on improving the process for making #4 and #6 hoses, then tackle the process for making #8 and #10 hoses.

The team members meet with the value stream champion to show him the results of their analysis. The champion agrees with the product family the team has selected, and the team records the product family in the value stream box on the storyboard.

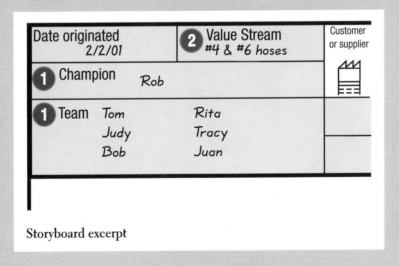

Storyboard excerpt

3. Learn about Lean

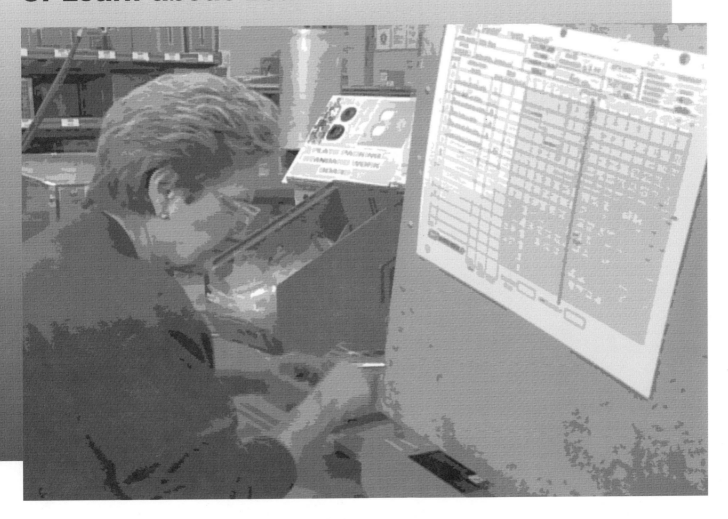

Step 3. Learn about Lean

In Steps 1 and 2 of Value Stream Management, you gained management commitment, formed a core implementation team, and identified the value stream targeted for lean conversion. However, before you can map the current state (Step 4), determine lean metrics (Step 5), and plan the future state (Step 6), you must gain a firm understanding of lean concepts. The purpose of Step 3 is to ensure that everyone has this understanding. After you learn about lean, you will start applying your knowledge by identifying non-lean conditions in the current state and entering them on the storyboard.

This step covers some key points on how to approach training and reviews the lean concepts that should be conveyed during training. As you read about each concept or tool, keep in mind that this is but one avenue or approach to learning about lean. The learning and implementation process differs for every organization. Integrate what makes good business sense for your organization. The bibliography includes a good list of resources that provide more detailed information on the lean concepts and tools discussed in this chapter.

Training and Doing—The Balancing Act

There is a delicate balance between training and doing. Lean concepts must make sense *now*, before you proceed with the next step. But ideally, you want to utilize the LEAP approach to all training:

LEarn . . . and then . . . **AP**ply

The faster *AP* follows *LE*, the better the results. You will learn much once you start creating and implementing. But if people are not asking questions, or if they seem disinterested, reconsider the approach immediately, before you go any further. More training or explanation on lean concepts and tools may be necessary.

Remember that the goal of learning is to get to the *doing* in Step 8. The point of a lean manufacturing transformation is to drive waste from the value stream. If you put a great deal of effort into planning a transformation that never actually occurs, all the time and effort that went into that process gets wasted.

Learning to Ride

Learning the Toyota Production System is a lot like learning to ride a bicycle. It cannot be done entirely in a classroom or by reading a book. You can draw a bicycle on a white board and tell people where to place their hands, where to sit, and where to put their feet. But this explanation is not enough to teach them how to ride. To become proficient at riding a bike, you must *learn by doing*. You may need some assistance at first, because you haven't yet learned how to balance. After several attempts and falls, however, you start to develop a sense or feel for riding.

Understanding and implementing the Toyota Production System is very similar. You can read material about the system and attend workshops and conferences on the subject. This will help, but like riding a bicycle, you must learn by doing. You will need some assistance in the beginning from people who have implemented lean before. They can help you keep your balance. As you begin to change your workplace with kaizen events, you will make mistakes. Don't give up when that happens. Pick yourself up and try again.

The Training Plan

All companies aspiring to become lean must place a premium on education and training. To get the core implementation team up to speed, develop a training plan based on the following five steps:

1. Determine the required skills and knowledge.

2. Assess current skill and knowledge levels of team members.

3. Determine the gap between present skills and knowledge and required skills and knowledge.

4. Schedule the training.

5. Evaluate the effectiveness of the training.

 Be sure to document the plan, making a specific agenda of activities, listing who will participate, and setting target completion dates.

The knowledge for the training should come from a variety of sources. Some good options for training include:

→ Conducting a simulation that ties together all the lean concepts. This can be accomplished by attending a public workshop, or by using materials your training department may supply.

→ Benchmarking another facility that is using some of the tools (see sidebar).

→ Demonstrating a successful in-house project.

→ Using internal resources to conduct just-in-time lean training sessions that are quickly followed by actual application of the concepts.

→ Using a consultant to facilitate the learning as it relates to the value stream.

→ Using books and videos combined with group discussion of the content.

The more you learn and do regarding lean, the more you *will be able to* learn and do. As with everything, true learning is cumulative; the way to gain experience with lean manufacturing is progressively, step by step. Build from what works and move on.

Benchmarking

Benchmarking is a structured approach to identifying a world-class process, then gathering relevant information and applying it within your own organization to improve a similar process.

Benchmarking Guidelines

❑ **Be specific in defining what you want to improve.** You may want to improve your entire manufacturing organization, but you also may want to see specifically how a company uses supermarkets and kanbans.

❑ **Be willing to share.** Identify an area you think may be world class in your organization, if you can, and present that to the potential benchmark site as something you are willing to share with them.

❑ **Attempt to make it a win-win experience.** Identify what's in it for the benchmark company! Offer something. Let them know that you are sincere.

❑ **Know the site.** Ensure that the benchmark team is familiar with some aspects of the company you will benchmark (products, size, whether it's a union shop, etc.).

❑ **Send questions.** Fax or e-mail specific questions in advance to the benchmark company's point person.

❑ **Don't go alone.** Do not benchmark in isolation. It is always better to have a minimum of two members on the benchmarking team.

❑ **Document.** Document and take notes as needed.

❑ **Respect privacy.** If some information is proprietary and cannot be released, respect that and move on.

❑ **Dress appropriately.** Be sure to discuss attire prior to the visit. Most companies have a "business-casual" dress code, but make sure you never underdress.

❑ **You can call.** Consider a conference call if an on-site visit is not practical.

❑ **Say "thanks!"** Show appreciation to the benchmark company. Consider offering appropriate gifts (i.e., t-shirts, hats, golf balls) to the people you will be visiting.

❑ **Follow up.** Follow up with a letter to the host facility detailing what you found helpful. Again, offer to be a benchmark site for them at any time in the future.

Key Concepts of Lean

What should be covered in the training? What are the key concepts people will need to be aware of as they work through the Value Stream Management process? The remainder of this chapter provides a brief overview of the concepts and tools people need in order to assess the current state and plan the future state effectively:

- The cost reduction principle.
- The seven deadly wastes.
- Two pillars of TPS:
  – JIT (just-in-time) production.
  – Jidoka (also known as autonomation).
- The 5S System.
- Visual workplace.
- Three stages of lean application:
  – Demand.
  – Flow.
  – Leveling.

The Cost Reduction Principle

Management is constantly under pressure from customers to reduce costs and lead times and to maintain the highest quality. Traditional thinking dictates that you set your sale price by calculating your cost and adding on a margin of profit. But in today's economic environment this is a problem. The market is so competitive that there is always someone ready to take your place. The customer can often set the price, and you don't have the luxury of adding a margin of profit.

Under these circumstances, the only way to remain profitable is to eliminate waste from your value stream, thereby reducing costs. This is the *cost reduction principle.* Determine the price customers are willing to pay, and subtract your cost to determine what your profit will be (profit = price – cost) (see Figure 3-1). Not only do customers often set the price, but also they often demand price reductions. This is why eliminating waste is so important—it's the primary means of maximizing profits.

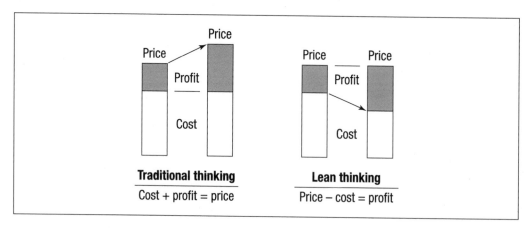

Figure 3-1. Cost plus versus price minus

Implementing lean has become a survival strategy in a manufacturing environment where mandatory cost reductions are a fact of life. An organization's resources should be focused on installing the proper systems to achieve cost reductions and the highest standards for quality and on-time delivery. Value Stream Management will allow you to deliver those results, provided you have a plan to ensure that resources are committed in the right places at the right times.

The Seven Deadly Wastes

The ultimate lean target is the total elimination of waste. Waste, or *muda*, is anything that adds cost or time without adding value. Over the years, seven deadly wastes have been identified (see Figure 3-2):

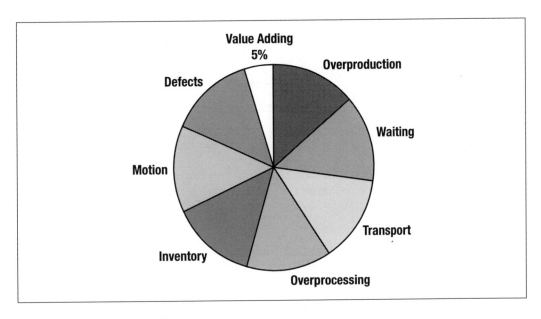

Figure 3-2. The seven deadly wastes

1. Waste of *overproducing*: producing components that are not intended for immediate use or sale.

2. Waste of *waiting*: idle time between operations or during an operation due to missing material, an unbalanced line, scheduling mistakes, etc.

3. Waste of *transport*: moving material more than necessary. This is often caused by poor layout.

4. Waste of *processing*: doing more to the product than necessary. This is the single most difficult type of waste to identify and eliminate. Reducing such waste often involves eliminating unnecessary work elements (including inspection through implementation of jidoka).

5. Waste of *inventory*: excess stock in the form of raw materials, work-in-process, and finished goods.

6. Waste of *motion*: any motion that is not necessary to the successful completion of an operation. Obvious forms of motion waste include back-and-forth movement in a workstation and searching for parts or tools. A more subtle form of motion waste involves any change in a worker's center of gravity. Thus, any time a worker stretches, bends, or twists, it is a waste of motion.

7. Waste of *defects and spoilage*: producing defective goods or mishandling materials. This includes the waste inherent in having to rework parts not made correctly the first time through. It also includes productivity losses associated with disrupting the continuity of a process to deal with defects or rework.

Within these seven categories are many more specific types of waste. To further define waste and understand how to address it, it is helpful to think of three different levels. Level one is gross waste, or low-hanging fruit. Specific wastes at this level are relatively easy to spot, and dealing with them can have a big impact. Level two is process and method waste, and level three is microwaste within processes. Clear away the lower-level wastes first to expose the higher-level wastes.

LEVELS OF WASTE		
Level One **Gross waste**	**Level Two** **Process and method waste**	**Level Three** **Microwaste within process**

• Work-in-progress
  – Poor plant layout
  – Rejects
  – Returns
  – Rework
  – Damaged product
  – Container size
  – Batch size
  – Poor lighting
  – Dirty equipment
  – Material not delivered to point of use
	• Long changeover
	  – Poor workplace design
	  – No maintenance
	  – Temporary storage
	  – Equipment problems
	  – Unsafe method
		• Bending and reaching
		  – Double handling
		  – Excess walking
		  – Look for stock
		  – Paperwork
		  – Speed and feed
		  – No SOP

The Two Pillars of the Toyota Production System

The most highly developed lean system in existence is the Toyota Production System. In fact, the terms "lean manufacturing" and "Toyota Production System" are interchangeable; and most if not all of the lean concepts discussed in this book were perfected at Toyota.

Two pillars support the Toyota Production System (see Figure 3-3):

- Just-in-time (JIT) production—the ideal state of continuous flow characterized by the ability to replenish a single part that has been "pulled" by the customer.
- Jidoka (autonomation)—the practical use of automation to mistake-proof the detection of defects and free up workers to perform multiple tasks within work cells.

The foundation on which these pillars rest is *people*, and the critical role they play in eliminating waste from manufacturing and business processes.

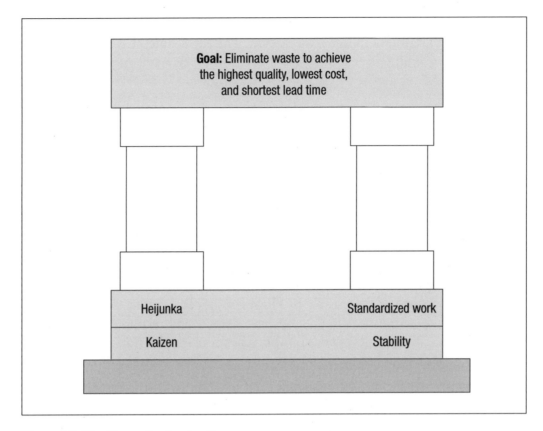

Figure 3-3. The Toyota Production System

People and Kaizen

At Toyota, employees are encouraged to make positive contributions toward improving their own work areas. Through kaizen events, teams meet for a short period to analyze conditions, recommend improvements, and implement them. The word *kaizen* comes from the Japanese characters "kai," or take apart, and "zen," or make good. Toyota is best known for the many informal kaizen ideas generated each day. As people at your site learn about and apply lean tools and concepts, their increased knowledge and awareness will result in increasing returns in the ongoing effort to eliminate waste from your manufacturing and business processes.

JIT (*Continuous Flow Production*)

The first Toyota Production System pillar represents just-in-time (JIT) production. JIT is synonymous with continuous flow production, the goal of which is to provide every customer with the highest quality products while meeting highly specific order and delivery requirements:

- *only* those units ordered;
- *just when* they are needed; and
- in the *exact amount* needed.

This encompasses not only finished goods, but all material delivered to the next user or "internal customer" throughout the value stream. The ideal state of continuous flow is characterized by the ability to replenish a single part that has been "pulled" by the customer. This ideal state is also referred to as *one-piece flow*.

For JIT to function seamlessly, tools such as value stream mapping, takt time, standardized work, kanban, and a supermarket pull system must be present. This chapter describes each method, roughly in their order of use.

Jidoka (Autonomation)

The second pillar of the Toyota Production System is jidoka, sometimes referred to as "autonomation," or automation with a "human touch." Jidoka means the practical use of automation to mistake-proof the detection of defects and free up workers to perform multiple tasks within work cells. The goal of jidoka is zero defects—to never pass a defective product downstream, and to eliminate the risk that an undetected defect will end up in the hands of the customer. However, jidoka is more than just zero defects. The purpose of jidoka is to achieve zero defects and promote flow within a JIT system. Every step and every task of improvement using the jidoka principle accomplishes both goals.

Jidoka is accomplished slowly, systematically, and inexpensively. It ensures that machines do only value-adding work. Implementing jidoka reduces cycle time and prevents wastes such as waiting, transport, inspection, and of course, defects. Moreover, jidoka can be applied to virtually any production process you have created.

The Three Functions of Jidoka

1. Separate human work from machine work.

2. Develop defect-prevention devices.

3. Apply jidoka to assembly operations.

Jidoka uses automation in such a way as to promote *flow*. By contrast, in traditional manufacturing operations, automated equipment has done little to promote the flow of goods. Instead, manufacturers often get stuck with extremely expensive, sophisticated, and *unreliable* equipment that will not operate continuously and make quality parts consistently. What's worse, such equipment often has been installed to improve a specific operation rather than a process. To improve flow, you must consider how the parts of the process relate to the whole and use automation *judiciously* to achieve your objectives.

Understanding the principles of jidoka will help you understand how to use automation to promote a smooth, defect-free process flow. The concepts and tools behind *poka-yoke* (mistake-proofing) and the statistical methodologies of Six Sigma will assist in this area immensely.

The 5S System—A Key Prerequisite for Lean

The 5S system is designed for organization and standardization of any workplace, including offices. It is a prerequisite to the implementation of any other improvement method. By implementing 5S , you will:

✓ Teach everyone the basic principles of improvement.

✓ Provide a starting place for eliminating all waste.

✓ Remove many obstacles to improvement (with very little cost).

✓ Give workers control over their workplace.

The 5S system consists of five activities:

Sort—sorting through the contents of an area and removing unnecessary items.

Set in Order—arranging necessary items for easy and efficient access, and keeping them that way.

Shine—cleaning everything, keeping it clean, and using cleaning as a way to ensure that your area and equipment is maintained as it should be.

Standardize—creating guidelines for keeping the area organized, orderly, and clean, and making the standards visual and obvious.

Sustain—educating and communicating to ensure that everyone follows the 5S standards.

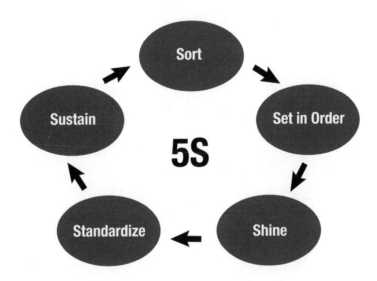

Although 5S activities seem like "good things to do," they aren't carried out for that reason. The 5S system is not merely housekeeping. It will have a positive impact on performance that will be reflected in the following metrics:

✓ Reduced total lead time.

✓ Elimination of accidents.

✓ Shorter changeover times.

✓ Improved worker attendance.

✓ Value-added activities.

✓ More improvement ideas per worker.

Visual Workplace

A visual workplace or visual factory begins with one simple premise: "One picture is worth a thousand words." If that picture is available exactly when you need it, where you need it, with just the right amount of information, then it's worth several thousand words. For that reason, the essence of the visual factory is "just-in-time information."

On the shop floor, the goal of a visual factory is to give people control over the workplace. There are several levels of control that apply (see Figure 3-4).

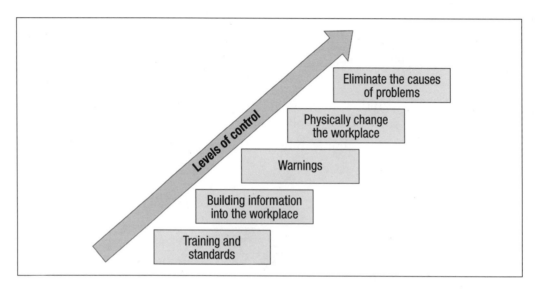

Figure 3-4. Levels of control

The visual factory is part of everything you do in lean. Keep this concept in mind as you learn about other lean concepts.

Three Stages of Lean Application: Demand, Flow, and Leveling

It is helpful to group lean concepts into three stages: demand, flow, and leveling. We will clearly define the tools and techniques within each stage, then outline a logical methodology for applying lean to the value stream.

1. Customer demand stage—understanding customer demand for your products, including quality characteristics, lead time, and price.

2. Flow stage—implementing continuous flow manufacturing throughout your plant so that both internal and external customers receive the right product, at the right time, in the right quantity.

3. Leveling stage—distributing work evenly, by volume and variety, to reduce inventory and WIP and to allow smaller orders by the customer.

We recommend that you implement these stages in the same order as we examine them here. One of the main reasons why lean transformations fail to be sustained is that people "cherry pick" their implementation tools-including popular kaizen and value stream mapping workshops. Understanding the demand, flow, and leveling stages of application, along with the guidelines for implementing VSM, will give you the solid approach required not only for implementing, but also for sustaining, lean improvements.

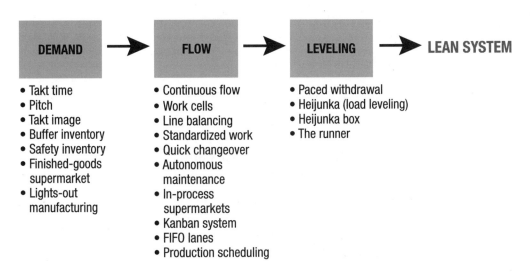

Think in terms of these three stages of demand, flow, and leveling and conduct kaizen events focused in each of these areas. As you proceed through the stages, the common principles or goals are to:

❑ **Stabilize** your processes, reviewing customer demand, equipment capabilities, labor balance, and material flow.

❑ **Standardize** your processes and the work.

❑ **Simplify** through kaizen, *after* you have stabilized and standardized processes.

Demand Stage

The various tools and concepts for determining and meeting demand include

- Takt time.
- Pitch.
- Takt image.
- Buffer and safety inventories.
- Finished-goods supermarket.
- Lights-out manufacturing.

Understand Customer Ordering Patterns

The first and primary concept of lean is to determine what you need to produce specifically in terms of quantity and delivery requirements; in other words, determine the actual number of parts you need to produce each day. This is not just a calculation; it means understanding your customers' ordering patterns. There are many sources for this information, including:

→ Sales forecasts.

→ Previous three months' actual production.

→ Current production forecasts.

→ Long-term agreements.

→ Interviews with customers.

Even though it may be difficult to determine customer demand, you can make a good start by finding out what you were producing last week and what you are producing this week.

Two Attitudes Toward Customer Demand

There are two opposite attitudes that influence your ability to meet customer demand: "There is no tomorrow!" or "There is always tomorrow!"

Meeting customer demand takes more than just applying the appropriate tools. You must take the attitude that there is no tomorrow—that you have only today to meet your customers' expectations.

The alternative is the attitude that prevails in many organizations today—the attitude that "we can get most of it done today, and finish it tomorrow." Such an attitude results in organizations achieving no better than 90 percent on-time delivery.

Which attitude will prevail at your factory?

Takt Time

From the data you collect on customer demand, you will determine your *takt time*, or the pace of customer demand. "Takt" is a German word for a musical beat or rhythm. Just as a metronome keeps the beat for music, takt time keeps the beat for customer demand. Takt time is the rate at which a company must produce a product to satisfy customer demand. Producing to takt means synchronizing the pace of production with the pace of sales.

To calculate takt time for a particular product family or value stream, divide the available production time by the total daily quantity required.

Takt time formula

$$\text{Takt time} = \frac{\text{Available production time}}{\text{Total daily quantity required}} \text{ or } \frac{\text{Time}}{\text{Volume}}$$

Note: Calculate takt time in seconds for high-volume value streams.

Let's say a manufacturing operation is open eight hours a day. To get the available production time, you must subtract from eight hours (the total available production time) the regularly scheduled planned downtime occurrences (for example, the time for lunch, breaks, or beginning-of-shift meetings).

Available production time calculation:

Available production time:	8 hours x 60 minutes	= 480 minutes
	minus two 10-minute breaks	= −20 minutes
	minus one 10-minute shift start meeting	= −10 minutes
	minus 30-minute lunch break	= −30 minutes
	480 − 60	= 420 minutes
To convert to seconds:	420 minutes X 60 seconds = 25,200 seconds	

The total of 25,200 seconds (420 minutes) is the available production time—the production time available for you to produce what the customer demands.

Let's say the customer demand is 420 parts per day. To calculate the takt time, divide the available production time (in seconds) by the total daily quantity required. The takt time would be 60 seconds, or one minute. This means that your processes must be set up to produce one unit every 60 seconds throughout the day. As order volume increases or decreases, takt time may be adjusted so that production and demand are synchronized.

Takt time calculation:

Takt time = available production time / total daily quantity required

Takt time = 25,200 seconds / 420 parts required = 60 seconds

Operational Takt Time

Another approach or adaptation of takt is a concept referred to as *operational takt time*. This is a time that is *faster* than takt time; it is used to balance the line to accommodate a chronic system failure such as equipment downtime, absenteeism, or a sudden customer demand change.

For example, if your takt time was 60 seconds but you knew of system problems that could affect production, you might try to work to an operational takt time that was 10 percent faster, or 54 seconds. This would help ensure that you could meet the true customer demand of the 60-second takt time.

Don't be satisfied with continually maintaining operational takt. Focus your kaizen activities to reduce system problems so you can move toward the 60-second takt that represents the true customer demand.

Pitch

The ideal state in any pull system is to eliminate all waste and create one-piece flow through the entire production system, from shipping back through raw material. However, we know that customers will not usually order product one piece at a time, but in a standard pack-out quantity shipped in a container of some sort. When this occurs, we must convert our takt time into a unit called *pitch*.

Pitch is the amount of time—based on takt—required for an upstream operation to release a predetermined pack-out quantity of work in process (WIP) to a downstream operation. Pitch is therefore the product of the takt time and the pack-out quantity.

Pitch formula:

Pitch = takt time × pack-out quantity

Note: Takt time is customer driven. Pack-out quantity may or may not be.

For example, if your takt time is 60 seconds and you want to move 20 pieces at a time, you would set a pitch of 20 minutes:

Pitch calculation:

Pitch = 60 seconds (takt time) × 20 pieces (pack-out quantity) = 1,200 seconds = 20 minutes

For high-volume, low-product-mix lines, pitch will normally be between 12 and 30 minutes, depending on customer requirements and any internal constraints.

Calculating pitch is a compromise between producing in large batches and implementing one-piece flow. For a variety of reasons, it is not always practical to produce to takt one piece at a time, but it is possible to produce a small batch of something in some multiple of the takt time based on the quantity produced. If your takt time is 0.5 seconds per part, for example, you are unlikely to achieve one-piece flow; you will have to settle for producing in small lots.

Advantages of Pitch

There are a number of advantages to producing in small batches based on pitch instead of producing large batches:

✓ A forklift is less necessary because you are working with smaller lots.

✓ Safety improves because workers are lifting smaller quantities.

✓ Inventory control improves.

✓ Problems can be identified immediately.

✓ You can react to a problem in a much shorter time than with large batches.

An advantage of working in pitch increments is that you can react to a problem in a much shorter time than if you were working in large batches. Pitch allows you to release a predetermined, manageable amount of work to the floor to meet customer demand and ensure that problems are detected quickly. If, for some reason, parts are not available at the specified pitch increment, it's important to notify the supervisor or team leader so that corrective action can be taken.

Takt Image: Visualize One-Piece Flow

To maintain the true spirit of lean, you must believe in and strive for the ideal state of one-piece flow, and challenge each compromise you make for practical reasons. You must ensure that you are doing everything possible to continuously improve so that you can meet the expectations of this ideal state. This vision of the ideal state is called *takt image*.

Takt Image

Takt image is the *vision* of an ideal state in which you have eliminated waste and improved the performance of the value stream to the point that you have achieved one-piece flow based on takt time.

Takt image challenges the entire organization to reach a higher level. Let's say that you are trying to improve a five-step process with a 60-second takt time. To achieve continuous flow in such a process with multiple operations—regardless of whether you are producing piece-by-piece or in small batches—each operation must be completed in 60 seconds or less. If the last of five operations in the process is an assembly operation that cannot presently produce one unit in 60 seconds or less, your ability to meet customer demand could be compromised. However, a clear understanding of takt image will motivate everyone to make the improvements necessary to achieve a faster assembly cycle time. Without a clear takt image, you run the risk that people will develop the attitude that "there is always tomorrow."

Buffer and Safety Inventories

As soon as you have determined customer demand, you must make the commitment to meet it—now. You do not want to wait until the future state is completed, as that may take months. However, if you cannot confidently meet demand with current production processes, you can use the tools of *buffer and safety inventory*. These are temporary measures that allow you to meet demand while you are planning and implementing improvements.

Buffer inventory is used when customer demand suddenly increases and your production process is not capable of meeting a lower (faster) takt time. Safety inventory, on the other hand, protects you from internal problems (labor power issues, quality problems, equipment reliability problems, power outages, and the like) that could prevent you from meeting demand.

By establishing buffer and safety inventories, you can meet demand without having to schedule overtime sporadically. But remember that buffer and safety inventories are compromises on the journey to your ideal state. Excess inventories are waste. As customer

---

Buffer and Safety Inventories

Buffer Inventory   Finished goods available to meet customer demand when customer ordering patterns, or takt time, varies.

Safety Inventory   Finished goods available to meet customer demand when internal constraints of inefficiencies disrupt process flow.

Note: These inventories should be stored and tracked separately. They exist for two distinct reasons.

---

demand becomes more stable and you improve the reliability of your operations and processes, you should periodically review these inventories and minimize or eliminate them, if possible (see Figure 3-5).

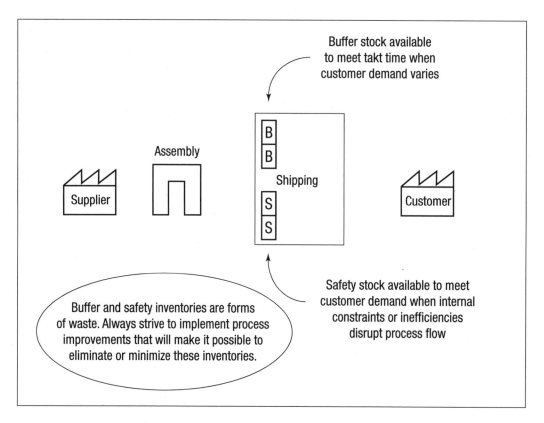

Figure 3-5. Buffer and safety inventories

Finished-Goods Supermarket

Next you must determine *where* customer demand is to be met within your value stream.

While shipping personnel are responsible for ensuring that products are shipped, *everyone* throughout the value stream is responsible for meeting customer demand. Shipping must be able to withdraw finished goods for shipment either from the end of the line or from a holding area or finished-goods warehouse.

We refer to such a holding area as a *finished-goods supermarket*—probably because the inspiration for Toyota's just-in-time pull system was the modern supermarket. Taiichi Ohno, who invented JIT, was fascinated by the idea that the physical flow of product culminated with its placement on supermarket shelves. He observed that once customers picked product off the shelves, the grocer replenished inventory by *pulling* from suppliers exactly what was needed to replenish inventory.

Similarly, in a finished-goods supermarket, items are not replaced until they are removed; they are removed when a customer orders them. This is the beginning of a pull system in which items are replenished by upstream operations as they are removed from shipping or the finished-goods supermarket. We will explain this unique withdrawal and replenishment system in more detail later when we discuss continuous flow and kanbans.

Finished-Goods Supermarket

A system used in the shipping part of the value stream to store a set level of finished goods and replenish them as they are "pulled" to fulfill customer orders. Such a system is used when it is not possible to establish pure, continuous flow.

Note: the inventory level in the supermarket does not include buffer and safety inventories.

Supermarkets are not just for finished goods. They can also be used—and in fact may be required to store—work-in-process in other parts of the value stream. We will cover the use of in-process supermarkets in our discussion of the flow stage.

Lights-Out Manufacturing

Lights-out or unattended manufacturing can be considered a means of meeting customer demand. It is allowing an automated machine to run when the operator is not present. This will increase the amount of product that can be manufactured. It is a fairly new concept that can work successfully, but you must consider the following:

- ❏ *Determine process capability*. The process must have a demonstrated C_{pk} (process capability index) of 1.63, or even 2.0.
- ❏ *Review type of material*. Some materials must be monitored closely to ensure product/equipment reliability.
- ❏ *Review part complexity*. Parts that are extremely complex are not good candidates for lights-out machining.
- ❏ *Determine the appropriate lot size*. The lot size should be consistent with one-pitch increments.

You must also consider the potential problems associated with lights-out machining:

✗ Operators may be reluctant to let a machine run when they are not there.

✗ If a quality problem occurs, it may occur throughout the entire lot.

✗ Time may be required to inspect the output.

> ### *Lights-Out Manufacturing*
>
> Lights-out manufacturing or "unmanned" machining is a method for meeting customer demand by making it possible for automated machines to run unattended for stretches of time-breaks, lunch, etc. This increases available production time, thus increasing capacity.

Lights-out manufacturing can increase production, but you must weigh the advantages against the concerns and the costs involved in order to decide whether the strategy makes good business sense.

Flow Stage

Once you have stabilized demand and devised a system for ensuring that you can meet it, you will turn your attention toward establishing a *flow* to ensure that customers receive the right parts at the right time in the correct amounts. The tools and concepts necessary to establish flow include

- Continuous flow—one-for-one manufacturing.
- Work cells.
- Line balancing.
- Standardized work.
- Quick changeover.
- Autonomous maintenance.
- In-process supermarkets.
- Kanban system.
- First-in, first-out (FIFO) lanes.
- Production scheduling

Continuous Flow

Pull

Continuous flow can be summarized in a simple statement: "move one, make one" (or "move one small lot, make one small lot"). Understanding continuous flow is critical to the just-in-time philosophy of ensuring that an upstream operation never makes more than is required by a downstream operation, so that a value stream never produces more than a customer requires (Figure 3-6).

Continuous flow processing means producing or conveying products according to three key principles:

- Only what is needed,
- Just when it is needed,
- In the exact amount needed.

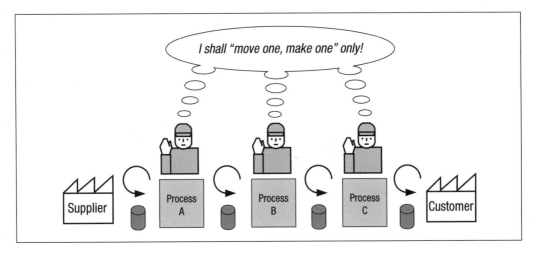

Figure 3-6. Continuous flow—"move one, make one"

One piece or one small batch is produced upstream only *after* a piece or a small batch is moved or "pulled" downstream. This is also called a *pull system* of production. Pull production is faster than batch or "push" production (see Figure 3-7). A pull system controls the flow between operations and eliminates the need for traditional production scheduling.

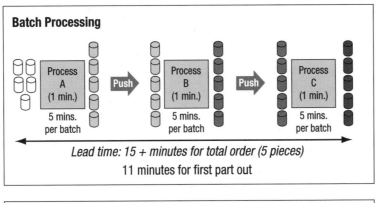

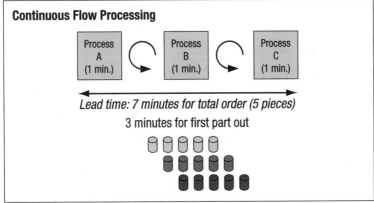

Figure 3-7. Push versus pull production system

Advantages of Continuous Flow

✓ Shorter lead times.

✓ Drastic reduction of work-in-process inventory.

✓ Ability to identify problems and fix them earlier.

✓ Makes traditional production scheduling obsolete.

It would be ideal to have continuous flow everywhere, but linking operations compounds problems each operation may have with:

✗ Lead times.

✗ Downtime.

✗ Changeover.

Other obstacles to continuous flow include poor plant layout and varying speed of processes.

➔ For example, consider a value stream consisting of four operations, each with a demonstrated 95 percent on-time delivery rate to its internal customer downstream. The cumulative effect would be an 81.4 percent on-time delivery rate (see Figure 3-8). If any of these operations were to have additional problems that further diminished the ability to achieve 95 percent internal on-time delivery, the overall loss would be even more dramatic. Until the processes can be improved, it may be necessary to use safety inventory between them to ensure that customer demand can be met on time.

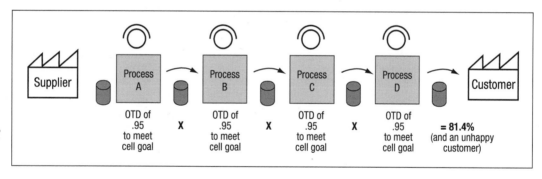

Figure 3-8. A flow system may compound problems unless processes are improved

Work Cells

U-Shaped Cell

In a flow system, production items must progress piece by piece (or in small batches) through the entire manufacturing process. Equipment must *not* be grouped by categories such as stamping, welding, grinding, machining, painting, etc., but in a way that minimizes transport waste and sustains continuous flow.

One way to achieve flow is to reconfigure operations into work cells. A work cell is a self-contained unit that includes several value-adding operations. The cell arranges equipment and personnel in process sequence and includes all the operations necessary to complete a product or a major production sequence. When operations are arranged into cells, operators can produce and transfer parts one piece at a time with improved safety and reduced effort.

Some principles to follow in planning cell layout include the following:

❏ Arrange processes sequentially.

❏ Set the cell up for counterclockwise flow (promotes use of the right hand for activities while the worker moves through the cell).

❏ Position machines close together, while taking safety into consideration for material and hand movement within a smaller area.

❏ Place the last operation close to the first.

❏ Create U- or C-shaped, or even L-, S-, or V-shaped cells, depending on equipment, constraints, and resource availability.

Keep the product demand and mix in mind when designing the cell layout. The cell must be able to adapt to customers' changing demands (Figure 3-9).

Line Balancing

Operator

Typically, some operations take longer than others, leaving operators with nothing to do while they wait for the next part. On the other hand, some operations may need more than one operator. Line balancing is the process through which you evenly distribute the work elements within a value stream in order to meet takt time. Line balancing helps optimize the use of personnel; it balances workloads so that no one is doing too little or too much. Keep in mind that customer demand may fluctuate, and changes in takt time often make it necessary to rebalance a line.

Line balancing begins with an analysis of your current state. The best tool to perform this task is an operator balance chart. The operator balance chart is a visual display of the work elements, time requirements, and operators at each workstation. It is used to show improvement opportunities by visually displaying each operation's times in relation to takt time and total cycle time.

Cycle Time

Cycle time is the time that elapses from the beginning of an operation until its completion— in other words, it is the processing time.

Don't confuse this measure of processing time with takt time, which is the measure of customer demand.

Total cycle time is the total of the cycle times for each individual operation in a value stream.

This is also referred to as *total value-adding time (VAT)*, because this is the time during which value is actually being added to the material as it flows through the process.

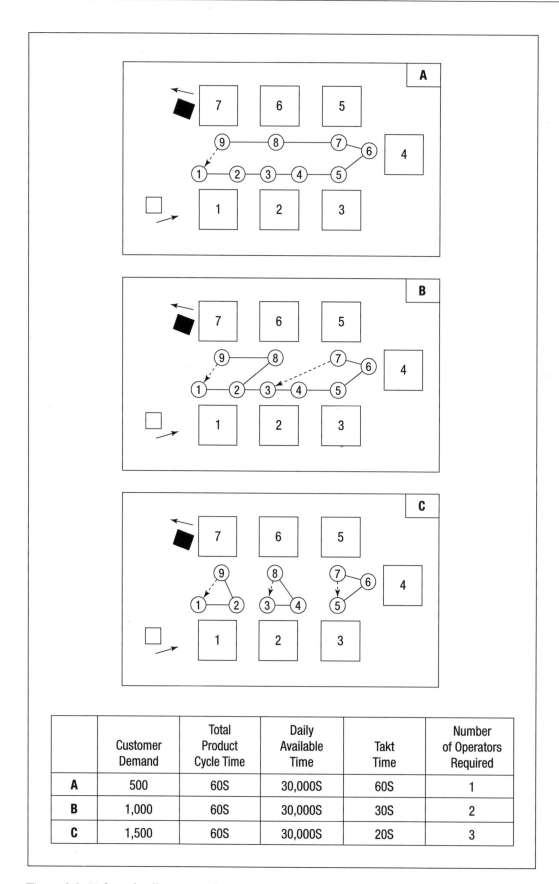

	Customer Demand	Total Product Cycle Time	Daily Available Time	Takt Time	Number of Operators Required
A	500	60S	30,000S	60S	1
B	1,000	60S	30,000S	30S	2
C	1,500	60S	30,000S	20S	3

Figure 3-9. U-shaped cell increases flexibility

The steps for creating an operator balance chart follow:

1. Determine current cycle times and work element assignments. For example consider the following process, which has five operations (A-E), four operators, a takt time of 60 seconds, and a total cycle time of 202 seconds:

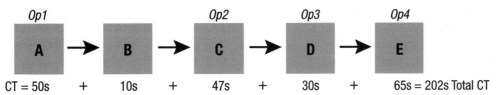

Takt time = 60 seconds

2. Create a bar chart that gives a better visual representation of the condition (see Figure 3-10). This current-state bar chart clearly shows a line that is out of balance, and where the imbalance exists.

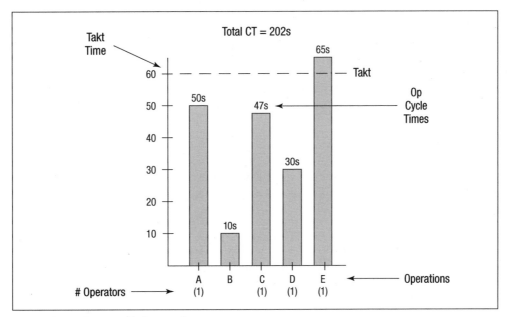

Figure 3-10. Operator balance chart—current state

3. Determine the number of operators needed by dividing total product cycle time by takt time:

$$\text{\# Operators needed} = \frac{202 \text{ (TCT)}}{60 \text{ (takt time)}} = 3.36$$

A requirement of 3.36 means that you do not really have enough work to keep four workers busy, but there is presently more work than three people can handle. This fact represents a problem, of course. But it also presents an opportunity to design an improved future state.

If you can eliminate enough waste in the process so that only three operators are required, you will reduce your direct labor cost per part and be able to re-deploy the fourth worker elsewhere. The conventions of lean thinking suggest that a decimal less than or equal to 0.5 (in this case, 0.36) is a good indicator that this is a realistic goal. In the improved process, each of the three remaining operators must do their share of what is necessary to make one part within the 60-second takt time (or a small batch within the time calculated for *pitch*). Thus, total cycle time must be less than or equal to 180 seconds.

One solution would be to combine operations A and B, and C and D, simplifying the new, combined operations so that one person can perform each of the three subprocesses (A-B, C-D, and E) in 60 seconds or less (see Figure 3-11).

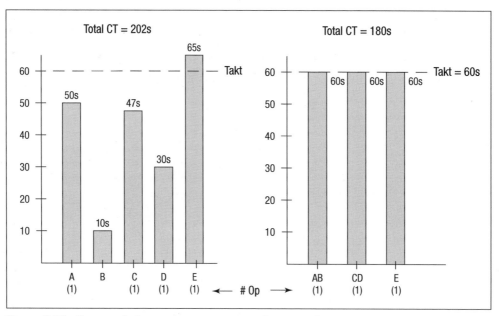

Figure 3-11. Operator balance chart—current and proposed

Now you have a target. However, you are probably wondering, "How are three operators going to complete their work within 60 seconds?" The best way to find the answer to this question is by implementing standardized work.

Standardized Work

For consistent flow to occur within the manufacturing value stream, workers must be able to produce to takt time and achieve consistent cycle times for the work elements assigned. You do not want one individual achieving a 45-second cycle time and a coworker achieving a 60-second cycle time for the same operation. You want to standardize to the 45-second cycle time and see to it that everyone does the same work the same way. This is accomplished by implementing standardized work.

Standardized work is an agreed-upon set of work procedures that establishes the best method and sequence for each manufacturing and assembly process. You can use a Standard Worksheet to illustrate the sequence of operations within a process, including operation cycle times (see Figure 3-12). This worksheet should be posted in the work area.

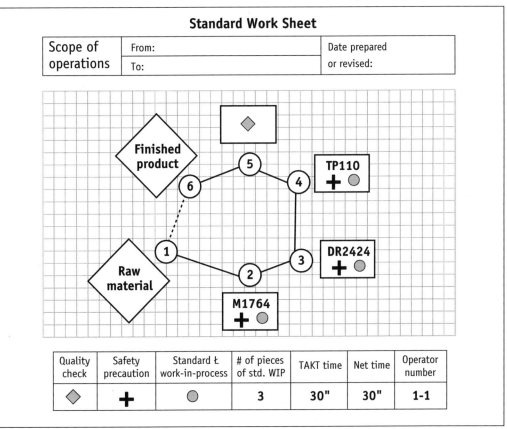

Figure 3-12. Standard Worksheet

Standardized work provides a basis for consistently high levels of productivity, quality, and safety. Employees develop kaizen ideas to continually improve these three areas. Here are some guidelines for implementing standardized work:

❏ Work together with operators to determine the most efficient work methods and ensure that consensus is attained. This may include reviewing the proposed set of revised work elements with the entire group that will be using them. Do not surprise people by unilaterally imposing new standards and procedures.

❏ Use the Standard Work Combination Sheet (Figure 3-13) to understand how process cycle time compares with takt time. This document displays the material and human workflow for a process. It specifies the exact time for each sequence within an operation, including walk time. If cycle time is longer than takt time, the operation can be "kaizened" (improved) to meet takt. This may include allocating some of the work elements into an operation that cycles faster than takt.

❏ Adhere to takt time, a critical unit of measurement for standardized work. Do not attempt to accommodate changes in takt time by making substantial changes in individual workloads. When takt time decreases, streamline the work and add employees as necessary. When takt time increases, assign fewer employees to the process.

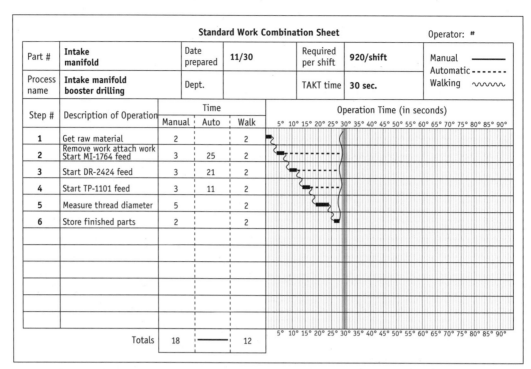

Figure 3-13. Standard Work Combination Sheet

Quick Changeover

When you establish takt time, create work cells, and implement standardized work, it is likely that you will also want to increase the variety of products flowing through the cells. Such flexibility requires tooling changes that do not disrupt continuous flow. The means for achieving this goal is the quick changeover (QCO) method. QCO originates from a methodology called single-minute exchange of die (SMED) that was developed by Shigeo Shingo at Toyota.

When to Implement QCO

The need for QCO usually becomes obvious at one of two stages—or at both of the following stages:

1. Demand stage: slow changeover times present a major obstacle to meeting customer demand.

2. Continuous flow stage: implementing standardized work underscores the need for faster changeover times to reduce total cycle time and help balance operations.

SMED is a theory and set of techniques that makes it possible to set up or change over equipment in less than 10 minutes. SMED begins with a thorough analysis of current setup procedures. It is applied in three sequential stages:

1. Distinguish between *internal setup* tasks that can be performed only while the machine is shut down and *external* setup tasks that can be performed while the machine is running.

2. Convert internal tasks to external tasks when possible; improve storage and management of parts and tools to streamline external setup operations.

3. Streamline all setup activities by implementing parallel operations (dividing the work between two or more people), using functional clamping methods instead of bolts, eliminating adjustments, and mechanizing when necessary.

Merely addressing the obvious things, like preparing and transporting tools and equipment while the machine is still running, can often cut setup time by up to 50 percent.

Autonomous Maintenance

Autonomous maintenance is a basic element of total productive maintenance (TPM). You can often prevent equipment-related losses such as breakdowns, speed losses, and quality defects by addressing the abnormal conditions that lead to such losses: inadequate lubrication, excessive wear due to contamination from grime or the by-products of production, loose or missing bolts, and so on.

Autonomous maintenance focuses on maintaining optimal conditions to prevent such losses. Autonomous maintenance has proved especially effective at reducing breakdowns and quality problems that disrupt continuous flow.

Autonomous Maintenance Steps

1. Clean and inspect equipment.

2. Eliminate sources of contamination.

3. Lubricate components and establish standards for cleaning and lubrication.

4. Train operators in general inspection of subsystems (hydraulics, pneumatics, electrical, etc.).

5. Conduct regular general inspections.

6. Establish workplace management and control.

7. Perform advanced improvement activities.

In-Process Supermarkets

Supermarket

Supermarket
Parts

Where obstacles to continuous flow exist, you can use an in-process supermarket system. A supermarket of work-in-process may be necessary to ensure that flow is possible. It is used when there are multiple demands made on a machine or a process.

Toyota found the supermarket to be the best alternative for scheduling upstream processes that cannot flow continuously. As you improve flow, the need for supermarkets may decrease. Remember that supermarkets are a compromise to the ideal state, as are pitch, buffer inventory, and safety inventory. You will not achieve your ideal state overnight, but keep the takt image alive and continually work toward that ideal state.

The supermarket system works best when there is a high degree of commonality between parts. Refer to the PQ analysis or part-routing matrix you created in Step 2 to examine part families.

Kanban System

Kanban Post

Kanban is at the heart of a pull system. Kanbans are cards attached to containers that store standard lot sizes. When the inventory represented by that card is used, the card acts as a signal to indicate that more inventory is needed. In this way, inventory is provided only when needed, in the exact amounts needed.

The Origin of Kanban

In Japanese, *kanban* means "card," "billboard," or "sign." Kanban refers to the inventory control card used in a pull system. Kanban also is used synonymously to refer to the inventory control system developed for use within the Toyota Production System.

Kanbans manage the flow of material in and out of supermarkets, lines, and cells. They can also be used to regulate orders from the factory to suppliers (see Figure 3-14).

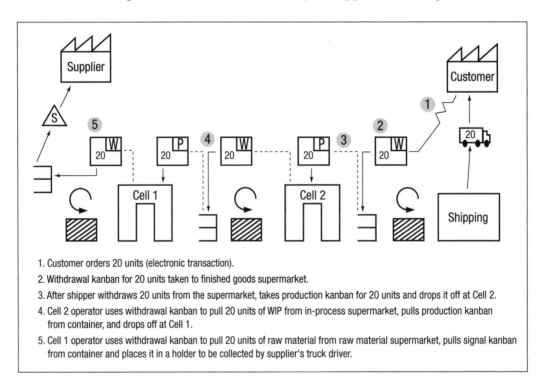

1. Customer orders 20 units (electronic transaction).
2. Withdrawal kanban for 20 units taken to finished goods supermarket.
3. After shipper withdraws 20 units from the supermarket, takes production kanban for 20 units and drops it off at Cell 2.
4. Cell 2 operator uses withdrawal kanban to pull 20 units of WIP from in-process supermarket, pulls production kanban from container, and drops off at Cell 1.
5. Cell 1 operator uses withdrawal kanban to pull 20 units of raw material from raw material supermarket, pulls signal kanban from container and places it in a holder to be collected by supplier's truck driver.

Figure 3-14. How a kanban system controls material flow

There are three types of kanban:

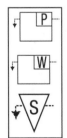

- A *production kanban* is a printed card indicating the number of parts that need to be processed to replenish what customers have pulled.
- A *withdrawal kanban* is a printed card indicating the number of parts to be removed from a supermarket and supplied downstream.
- A *signal kanban* is a printed card indicating the number of parts that need to be produced at a batch operation to replenish what has been pulled from the supermarket downstream.

You can think of kanbans as a factory's automatic nervous system. To work properly, a few rules must always be followed:

Kanban Rules

- ❏ Downstream operations or cells withdraw items from upstream operations or cells.
- ❏ Upstream operations or cells produce and convey only if a kanban card is present and only the number of parts indicated on the kanban.
- ❏ Upstream operations send only 100-percent defect-free products downstream.
- ❏ Kanban cards move with material to provide visual control.
- ❏ Continue to try to reduce the number of kanban cards in circulation to force improvements.

FIFO Lanes

If you lack a high degree of commonality between parts and cannot use an in-process supermarket system, then you can work with the concept of a first-in, first-out (FIFO) lane. First-in-first-out (FIFO) is an inventory control method used to ensure that the oldest inventory (first-in) is the first to be used (first-out). FIFO lanes are useful in situations where multiple value streams meet before product customization, and before large-batch operations where dissimilar parts go through an operation such as anodizing, welding, stamping, or painting (see Figure 3-15).

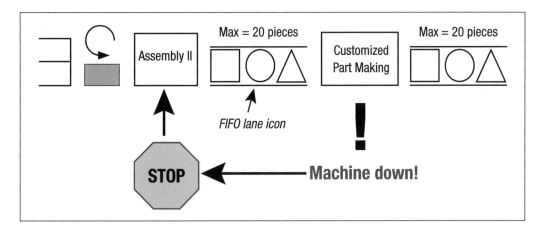

Figure 3-15. FIFO lanes

A FIFO lane has the following characteristics:

- ❏ Holds a designated number of parts between two processes and is sequentially loaded.
- ❏ Is created in such a way that it is difficult—if not impossible—to draw anything other than the oldest inventory first.
- ❏ Uses a signal to notify the upstream process to stop producing when the lane is full, preventing overproduction (one of the seven deadly wastes).

❏ Requires sequencing rules and procedures for upstream and downstream processes to ensure that neither overproduces and that time is not wasted.

❏ Requires discipline by the workforce to ensure FIFO integrity.

Note that even though in-process supermarkets and FIFO lanes are used in different situations, in both cases product should be pulled in FIFO *order*. Standardize processes for pulling product from in-process racks and for replenishing it so that FIFO order is maintained.

FIFO Lane Example

An upstream operation cuts steel rods for a cell that makes three slightly different but related products. One product starts with a 3-inch diameter rod that is 8 inches long, the second product starts with a 4-inch diameter rod that is 10 inches long, and the third product starts with a 5-inch diameter rod that is 12 inches long. Since the cell requires three different-sized rods, you could establish a FIFO lane to ensure that the rods are processed downstream in the order that they were cut.

On the other hand, if each of the three products were made from the same-sized steel rod, you could establish an in-process supermarket from which parts could be pulled when needed.

Production Scheduling

Remember that a pull system is based on actual need, not forecasts. Sustaining such a system requires you to schedule production and control inventory based on the needs of the downstream operation closest to the customer—and to use pull and continuous flow methods to initiate and signal all other activities. If necessary, you may have to introduce supermarkets and/or FIFO in the value stream to enable your production system to handle variations and mistakes. The scheduling point should be one that has zero downtime, little or no changeover (ideally less than one minute), and the greatest worker flexibility. Often this will be an assembly area.

Review of the Various Levels of Flow

When you plan the future state you will have to consider the best level of continuous flow for your value stream.

Flow in the Ideal State (One-Piece Flow)
Advantages:

✓ Absolute control over processes.

✓ Instant feedback on quality and safety issues.

✓ Balanced workload.

✓ Immediate reaction to system failure (machine, people, material).

✓ True takt image that can be clearly seen.

Disadvantages:

✗ None, assuming you have eliminated problems with downtimes, changeovers, etc.

Flow Using Supermarkets

Advantages:

✓ Allows for flow when using shared equipment.

✓ Better use of capital equipment when state-of-the-art technology has not risen to small, more lean-oriented machines.

✓ Potential for enhanced labor balance.

Disadvantages:

✗ Quality is harder to monitor and correct when producing in small batches.

✗ Erosion of takt image.

✗ Space required for storage.

✗ Loss of control.

Flow Using FIFO Lanes

Advantages:

✓ Allows for flow when potential for chronic failure of upstream process exists.

✓ Allows for flow during tool changes.

✓ Allows for complex labor demand within an assembly operation.

Disadvantages:

✗ Quality is harder to monitor and correct.

✗ Erosion of takt image.

✗ Space required for storage.

✗ Loss of control.

These examples represent only a few of the possible variations. You can develop an appropriate combined approach depending on your ability to stabilize and standardize machines, move material, and redeploy people within the value stream.

Leveling Stage

After you have determined demand and established flow, you will work on *leveling* production. Leveling involves evenly distributing over a shift or a day the work required to fulfill customer demand. The concepts and tools used to level production include

- Paced withdrawal.
- Heijunka.
- Heijunka box.
- The runner.

To maintain a takt image, you must establish a method by which you can balance the pace of production against the pace of sales or takt time. There are two ways to accomplish this: paced withdrawal and heijunka.

Paced Withdrawal

Paced Withdrawal

Paced withdrawal is a system for moving small batches of a product from one operation or process to the next, at time intervals equal to the pitch. Paced withdrawal is used when you have no product variety in the value stream, meaning that all pitch increments will be identical.

Remember that the ideal state is one-piece flow. However, customers usually want products in containers that hold a standard pack-out quantity. Paced withdrawal levels production by dividing the total requirement for a shift or day into batches equal to a pack-out quantity. The pitch determines the frequency with which containers are released to shipping.

Heijunka (Load Leveling)

Heijunka (Load Leveling)

Heijunka is a sophisticated method for planning and leveling customer demand by volume and variety over the span of a day or shift. If there is little or no product variation, you may not need this level of sophistication. As you move toward smaller lots or pure, continuous, one-piece flow, the demand for parts is subject to sudden peaks and valleys. Large orders may immediately deplete inventory, making it difficult to manage.

Heijunka may be the key to establishing a true lean pull system in your facility—if your product mix warrants it. Heijunka uses paced withdrawal based on pitch, but breaks it into units based on the volume and variety of product being manufactured. For example, consider a value stream that makes five related products in standard-pack quantities of 25, as shown in the table below.

Product	A	B	C	D	E
Daily requirement	300	200	200	50	50
Pack-out quantity	25	25	25	25	25
# of kanbans	12	8	8	2	2

In each case, the number of kanbans is determined by dividing the daily requirement by the pack-out quantity. The total daily requirement is 800 units, and the available production time over two shifts is 52,800 seconds. This means that takt time equals 66 seconds and pitch equals 1,650 seconds (27.5 minutes):

Takt time = Available production time / Total daily quantity required
Takt time = 52,800 seconds / 800 units
Takt time = 66 seconds

Pitch = Takt time X Pack-out quantity
Pitch = 66 seconds per unit X 25 units per container
Pitch = 1,650 seconds or 27.5 minutes

So, every 27.5 minutes, 25 units must be released to shipping. Now the question is: "25 units of which product?" Over the course of the day, the value stream must turn out 12 containers of A, 8 containers each of B and C, and 2 containers each of D and E. In other words, the ratio of A : B : C : D : E is 12 : 8 : 8 : 2 : 2. Reduced to its smallest terms, the ratio is 3 : 2 : 2 : 0.5 : 0.5—or, for every three containers of product A produced, two containers each of products B and C and 0.5 containers each of products D and E must be produced. The levelled production ratios are managed through distribution of kanbans using a heijunka box, explained in the next section.

Implementing heijunka clearly requires a sound understanding of customer demand and the effects of this demand upstream. Heijunka is not something that you can put in place with a cosmetic lean effort: it requires strict attention to the principles of stabilization and standardization.

Heijunka Box

The heijunka box, or leveling box, is a physical device used to manage levelled production volume and variety over a specified time period. The load is leveled with consideration for the most efficient use of people and equipment. Kanban cards are placed in slots corresponding to the pitch increments in which products are to be released to shipping and subsequently replenished.

In the heijunka example above, we determined that level production could be achieved based on the following ratio:

$A : B : C : D : E = 3 : 2 : 2 : 0.5 : 0.5$

Figure 3-16 represents a heijunka box. The left column shows 32 pitch increments, and the right column shows which product is staged for shipment and subsequently replenished during each 27.5-minute period.

Note that the heijunka box is loaded in a way that approximately reflects the ratio shown above:

- Product A is made during the first three pitch periods.
- Product B is made during the next two periods.
- Product C is made during the following two periods.
- Since production is based on pitch, it would be impractical to make a half container of product D. Instead, it would make good sense to run A-the high runner-for three more pitch periods before running D. This helps sustain flow by minimizing changeovers.
- After running D, you would dedicate the next four pitch periods to running two more containers of B and C respectively.
- Product E is made during the final pitch period of the first shift.
- The pattern established above repeats during the second shift.

Pitch Increment	Product
6:30 — 6:40	Beginning of 1st Shift Meeting and Operator PM Checks
6:40:00 — 7:07:30	A
7:07:30 — 7:35:00	A
7:35:00 — 8:02:30	A
8:02:30 — 8:30:00	B
8:30 — 8:40	Break—No Production
8:40:00 — 9:07:30	B
9:07:30 — 9:35:00	C
9:35:00 — 10:02:30	C
10:02:30 — 10:30:00	A
10:30 — 11:00	Lunch—No Production
11:00:00 — 11:27:30	A
11:27:30 — 11:55:00	A
11:55:00 — 12:22:30	D
12:22:30 — 12:50:00	B
12:50 — 1:00	Break—No Production
1:00:00 — 1:27:30	B
1:27:30 — 1:55:00	C
1:55:00 — 2:22:30	C
2:22:30 — 2:50:30	E
2:50:30 — 3:00:00	End of 1st Shift 5S
3:00 — 3:10	Beginning of 2nd Shift Meeting and Operator PM Checks
3:10:00 — 3:37:30	A
3:37:30 — 4:05:00	A
4:05:30 — 4:32:30	A
4:32:30 — 5:00:00	B
5:00 — 5:10	Break—No Production
5:10:00 — 5:37:30	B
5:37:30 — 6:05:00	C
6:05:30 — 6:32:30	C
6:32:30 — 7:00:00	D
7:00 — 7:30	Lunch—No Production
7:30:00 — 7:57:30	A
7:57:30 — 8:25:00	A
8:25:00 — 8:52:30	A
8:52:30 — 9:20:00	B
9:20 — 9:30	Break—No Production
9:30:00 — 9:57:30	B
9:57:30 — 10:25:00	C
10:25:00 — 10:52:30	C
10:52:30 — 11:20:00	E
11:20 — 11:30	End of 2nd Shift 5S

Figure 3-16. Heijunka box—pitch increment

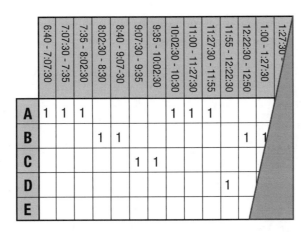

Figure 3-17. Heijunka box—kanbans

Note that by the end of the day, production requirements have been met. Twelve containers of *A*, eight containers each of *B* and *C*, and two containers each of *D* and *E* have been produced.

We created the table for the preceding example to show in a simple way on one page how the production of a variety of different products is distributed in a balanced way over an entire day. A heijunka scheme is more commonly represented as shown in Figure 3-17 (in reality, you might fabricate a box with slots in which you place kanbans or a board on which the kanbans are posted).

The Runner

In a lean transformation you often discover in the course of performing line balancing that you can eliminate a worker from a target value stream and redeploy that person elsewhere. Sometimes you can redeploy a displaced operator as a runner, or material handler, provided that he or she is qualified to assume that role (see "Runner Qualifications").

The runner ensures that pitch is maintained. He or she covers a designated route within the pitch period, picking up kanban cards, tooling, and components, and delivering them to their appropriate places. If a heijunka box is used, the runner removes kanbans from it to use as "visual" work orders. In a sense, the heijunka box is like a mailbox for the value stream, and the runner is the mailman. If a heijunka box is not being used,

---

Runner Qualifications

• Understands value stream production requirements.

• Communicates well.

• Able to recognize and report abnormalities.

• Understands lean concepts.

• Understands the importance of takt time and pitch.

• Works efficiently and precisely.

then the runner picks up and delivers parts from store locations as required to sustain efficient flow through the work areas or cells.

Runners play an important role in proactive problem solving. Because they continuously monitor the functioning of a line or cell as well as pitch (or takt time), runners are closely attuned to how well the value stream is fulfilling customer requirements. Normally, when a problem occurs, an operator immediately notifies a team leader or supervisor and the problem is addressed *after* it occurs. However, runners are in a unique position to help prevent small problems before they become large problems that seriously disrupt process flow.

Remember that leveling occurs *after* you have achieved continuous flow. It is a refinement of your lean design. You may find that specific techniques implemented earlier will be eliminated as you successfully level production.

Identify Non-Lean Conditions

After learning something about lean, you will begin to look at the value stream in a new way. As a first step in applying your knowledge, start observing ways in which the manufacturing process is not lean, or could be improved. Ask yourself questions about what you are observing, such as:

❑ Does the floor layout promote waste-free flow of parts through the process?

❑ Do you have a push system or a pull system? How do you communicate orders to the upstream process? How do you pass components to the downstream process?

❑ Are you producing in large lots, small batches, or one-piece flow?

❑ Is customer demand being met? Are you working to takt time? Have you determined the pitch?

❑ Is the work area messy and disorganized?

❑ Are you tolerating inventory, waiting, and other forms of waste?

❑ How long are your changeover times?

Enter your observations on the storyboard in the space labeled for Step 3.

PREMIERE MANUFACTURING CASE STUDY, STEP 3

The team creates a training plan and spends the next six weeks learning about lean manufacturing tools and methods.

Two of the members attend a one-day seminar on lean. On June 15, the entire team attends a four-hour overview on lean manufacturing techniques conducted by Premiere's training department. The overview includes a simulation that demonstrates the difference between batch production and continuous flow production. On June 30, five team members go on a benchmarking trip to a local company that has successfully implemented lean methods. By July 30, the team leader completes and gives a report on *Lean Thinking*, by Womack and Jones, and *The Toyota Production System*, by Taiichi Ohno.

Team Member	Training Activity	Completed by
Bob	Attend Overview/Simulation	6/15
	Benchmark Company C	6/30
	Attend Lean Workshop	7/15
Rob	Attend Overview/Simulation	6/15
	Attend Lean Workshop	7/15
Juan	Attend Overview/Simulation	6/15
	Benchmark Company C	7/15
Judy (Team Facilitator)	Attend Overview/Simulation	6/15
	Benchmark Company C	6/30
Rita	Attend Overview/Simulation	6/15
	Attend Lean Workshop	7/15
Tom (Team Leader)	Attend Overview/Simulation	6/15
	Benchmark Company C	6/30
	Read Lean Thinking	7/30
	Read Ohno's Toyota Production System	7/30
Tracy (Scrbe)	Attend Overview/Simulation	6/15
	Benchmark Company C	6/30

Premier Manufacturing Lean Training Plan

After discussing what they have learned and observed, the team members conclude that the target value stream is operating currently with a push system and limited continuous flow. The work areas are generally disorganized and disorderly, neither takt time nor pitch is being used, and there is tremendous variation in the way different operators perform value-adding tasks (in other words, it will be necessary to implement standardized work). The team enters these observations on the storyboard and looks forward to mapping the current state in Step 4.

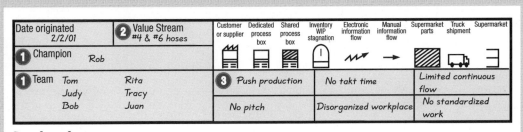

Storyboard excerpt

4. Map the Current State

Step 4. Map the Current State

After attaining a solid understanding of lean, the next step is to map the current state of production, showing the flow of material and information. With this step, you place a stake in the ground. Your goal is to gather *accurate, real-time* data related to the product family, or value stream, you specified in Step 2. Therefore, you must go to the factory floor to collect data rather than rely on past reports generated by an industrial or process engineer, or on data based on someone's best recollection. Moreover, data collection is not a solitary activity; it is important for the core team to work together.

Value Stream Mapping

In their book, *Lean Thinking*, James Womack and Daniel Jones refer to value stream mapping as identifying all the specific activities occurring along a value stream for a product or family. (See Figure 4-1.)

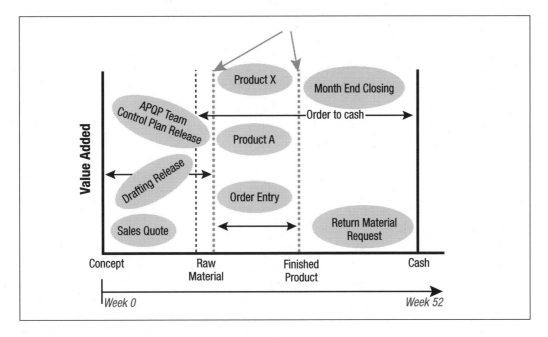

Figure 4-1. Typical manufacturing value stream focus

There are numerous ways to determine the scope of a value stream map. Here are a few common ones:

→ You can define activities and measure the time it takes to go from conceiving a product to launching it.

→ You can define the activities and measure the time it takes from receiving raw materials to shipping finished parts to a customer.

→ You can define the activities that take place from the time an order is placed until cash is received for the finished order.

In this book, we focus on defining the activities and measuring the time it takes from receiving raw materials to shipping finished parts to your customer or customers.

Map Material and Information Flow

Because a value stream map gives a visual representation of material and information flow for a product family (value stream), it is indispensable as a tool for visually managing process improvements. To improve a process you must first observe and understand it. Mapping a process gives you a clean picture of the wastes that inhibit flow. Eliminating waste makes it possible to reduce manufacturing lead time, which will help you consistently meet customer demand (Figure 4-2).

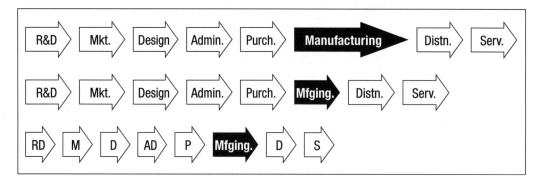

Figure 4-2. Lead-time reduction through waste elimination

As we delve more deeply into use of the value stream mapping tool, keep in mind that a key to establishing a lean *material flow* is understanding how *information* flows—that is, how production scheduling is achieved. As you gather data at each point on a specific value stream, keep asking, "How do you know what to make next?" This will allow you to trace information flow along with material flow on your map. Capturing this information is the essence of value stream mapping (Figure 4-3).

When you place your current-state map on the VSM storyboard, you are helping to promote good visual management on the factory floor. The current-state map represents your baseline. With just a bit of training on the meaning of the value stream mapping icons shown in Figure 4-4, anyone should be able to discern information on actual material and information flow.

Mapping material and information flow will allow you to:

✓ Visualize the entire manufacturing material and information flow, instead of a single, isolated operation (such as fabrication, welding, or assembly).

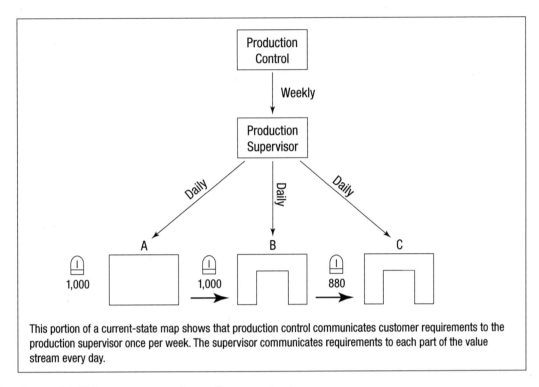

This portion of a current-state map shows that production control communicates customer requirements to the production supervisor once per week. The supervisor communicates requirements to each part of the value stream every day.

Figure 4-3. Value stream map: lines of communication

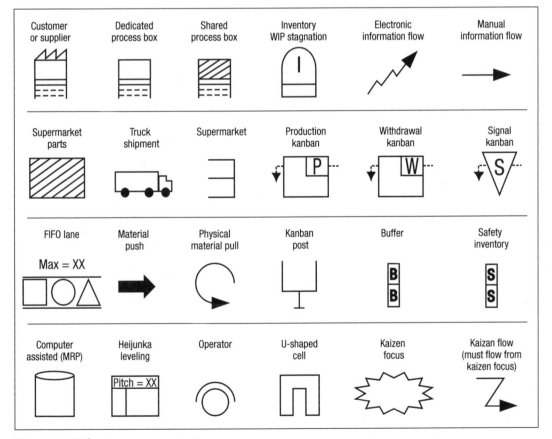

Figure 4-4. Value stream mapping icons

✓ Visualize how operations currently communicate with production control and with each other.

✓ See problem areas and sources of waste.

✓ Locate bottlenecks and WIP.

✓ Spot potential safety and equipment concerns.

✓ Provide a common language for all manufacturing personnel.

✓ Gain insight into how the operation truly is running that day.

How to Map the Current State

As we explore how to map the current state, we will alternate generic mapping procedures that will help you conduct your own current-state mapping with details from the Premiere Manufacturing Case Study.

Getting Ready

There are three major preparations for mapping the current state:

1. Working in a conference room as a team, draw rough sketches of the main production operations on a white board.

2. Go to the floor, beginning with the most downstream operation (i.e., shipping), and collect actual process data. Use the Attribute Collection Checklist to gain consensus on what data you need. The team should select 7 to 10 key attributes. Create a "parking lot" list of lean attributes that are important but not critical to the team's charter. While you are gathering data, take notes on information and material flow.

ATTRIBUTE COLLECTION CHECKLIST

☐ Total time per shift

☐ Regularly planned downtime occurences such as breaks and lunch that reduce available time

☐ Total available daily production time (subtract regularly planned downtime occurences from the total time per shift)

☐ Delivery schedules

☐ Number of parts per shipping container

☐ Quantity of parts shipped per month and per day (by part)

☐ Cycle times

☐ Changeover times

☐ Work-in-process (WIP) amounts

☐ Actual lot sizes

☐ Pitch increments (if available)

☐ Economic lot sizes (economic order-quantity)

☐ Number of operators

☐ Reliability metrics—expressd in terms of mean-time between failures, uptime. or overall equipment effectiveness (the product of availability, performance, and quality rates)

☐ Shifts on which the process operates

☐ Line speeds

☐ Preventive maintenance schedules (important, because they can reduce net available time for a particular process)

☐ Disruptions in manufacturing flow

☐ Exceptions that may occur due to rework

☐ _____

☐ _____

☐ _____

☐ _____

3. Regroup away from the floor to discuss the results of the data gathering efforts and make sure that all necessary data has been collected.

The Etiquette of Factory Floor Research

Whever you go out on the factory floor to gather data, be sure to do the following:

❏ Communicate to all areas you will be studying before going to the floor.

❏ Make proper introductions when you arrive on the floor. Workers like to know what's going on—especially if outside consultants or representatives from a supplier or customer are present.

❏ Explain your purpose for studying operations.

❏ Be open and honest in responding to questions or issues that arise.

❏ Respect people's workspace, and thank them for their contributions. Use the opportunity to reinforce mutual trust and respect.

PREMIERE MANUFACTURING CASE STUDY: STEP 4

In Step 2, the core implementation team at Premiere Manufacturing chose to focus on the value stream for the #4 (0.25-inch) and #6 (0.125-inch) coolant hoses. In Step 3, the team received a comprehensive overview of lean manufacturing concepts. Now the team is ready to prepare for mapping the current state.

The team convenes and sketches the main production processes for the # 4 and #6 hoses.

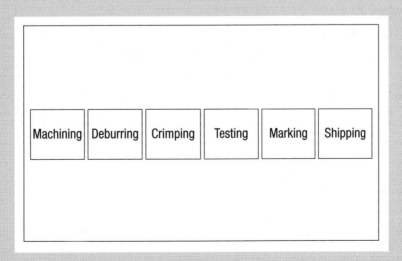

The team reviews the attribute checklist and decides to be sure to collect data on the following process attributes:

❏ Quantity of parts shipped per month and per day.

❏ Supplier delivery schedule.

❏ Available production time.

❏ Cycle time.

❏ Changeover time.

❏ Uptime.

❏ Number of operators.

❏ Number of shifts.

❏ Inventory locations and quantities.

❏ Time between processes.

The team goes to the floor to collect the data, beginning with the shipping area and working back to the machining process. They reconvene to review their notes before beginning to map the current state. The data they collected is shown under the heading "Premiere Process Attributes" in the form shown that follows.

Premiere Current State Data Collection

Customer Requirements

- Average demand: 10,080 units per month = 504 per day:
  - ✓ 6,720 #4 hoses   336 per day
  - ✓ 3,360 #6 hoses   168 per day

- Shipping month: 20 days
- Units per container: 24
- Containers per day: 21

Supplier Information

Premiere receives a weekly shipment of 2,500 units from its supplier, ABC, Inc.

Premiere Process Attributes

Availability: Total available production time is 8.5 hours (510 minutes) per shift. There is a 30-minute unpaid lunch break and two 10-minute breaks—a total of 50 regularly scheduled minutes of planned downtime. Therefore, the available production time is 460 minutes (27,600 seconds) per shift.

Shipping:
- Location = staging area
- Frequency/method = daily/UPS
- Finished-goods inventory = 2,000 units

Marking:
- Cycle time = 50 seconds
- Changeover = 5 minutes
- Availability = 27,600 seconds
- Uptime = 99%

Marking—continued:
- Operator = 1
- WIP = 2000 units between testing and marking
- Time between marking and shipping: 4 days

Testing:
- Cycle time = 30 seconds
- Changeover = 5 minutes
- Availability = 27,600 seconds
- Uptime = 99%
- Operator = 1
- WIP = 2000 units between crimping and testing
- Time between testing and marking: 4 days

Crimping:
- Cycle time = 40 seconds
- Changeover = 5 minutes
- Availability = 27,600 seconds
- Uptime = 99%
- Operator = 1
- Shifts = 1
- WIP = 3,500 units between deburring and crimping
- Time between crimping and testing = 4 days

Deburring:
- First-in, first-out (FIFO) lane between machining and deburring
- Cycle time = 5 seconds
- Changeover = 0
- Availability = 27,600 seconds
- Uptime = 100%
- Operator = 0 (Machining operator also runs deburring when necessary)
- Shifts = 1
- WIP = 3,360 #4 and 1,680 #6 between Machining and Deburring
- Time between deburring and crimping: 7 days

Machining:
- Cycle time (CT) = 45 seconds
- Changeover time (C/O) = 60 minutes
- Availability = 27,600 seconds
- Uptime = 87 percent
- Operator = 1
- Shifts = 1
- WIP: 2,500 prior to machining
- Time between machining and deburring: 10 days

Flow of Information and Material
- All communications with customer and supplier are electronic.
- Production control receives monthly forecasts and weekly orders from Cord.
- Production control transmits monthly forecasts and weekly orders to ABC (supplier).
- There is a weekly order release to the production supervisor.
- There is a daily order release to machining, crimping, testing, and marking.
- All material is pushed, so there is a push icon between each process.
- There is a FIFO lane between machining and deburring.

Creating the Current-State Map

Once you have compiled data, you are ready to map the current state. First, however, take some time to review the icons used in mapping the value stream (see Figure 4-4).

Use the following steps for creating the current-state map:

1. Draw icons representing the customer, supplier, and production control.

❑ Use the same icon to represent the customer and the supplier.

❑ Draw the *customer icon* in the upper right corner of the sheet.

❑ Draw the *supplier icon* in the upper left corner.

❑ Draw the *production control icon* between the customer and supplier icons.

2. Draw a *data box* below the customer icon, and enter the customer requirements in it. Include monthly and daily requirements of each product, and the number of containers required per day.

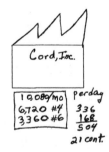

Calculating Daily Requirements

If demand is constant, calculating daily requirements is simple. If you use the customer requirement data from the case study cited earlier, you know that Premiere ships on average a total of 10,080 hoses per month to Cord, Inc.—6,720 #4 hoses and 3,360 #6 hoses. You also know that Premiere ships 20 days per month. Therefore, if you divide the requirements by 20, you can determine the average number of hoses shipped per day:

- 10,080 hoses per month (20 days = 504 hoses (#4 and #6 combined) per day

- 6,720 #4 hoses per month (20 days = 336 #4 hoses per day

- 3,360 #6 hoses per month (20 days = 168 #6 hoses per day

3. Enter shipping and receiving data:

❑ Draw a *truck icon* below the customer icon, and enter inside the truck icon the delivery frequency. (How often does the customer require shipments?)

❑ Draw a *shipping icon* below the customer truck.

❑ Draw a *direction arrow* from the shipping icon, through the customer truck (as if the arrow goes behind the truck), to the customer icon.

❑ Draw a *truck icon* below the supplier icon, and enter inside the truck icon the delivery frequency. (How often are raw materials required?)

❑ Draw a *direction arrow* from the supplier, through the supplier truck (as if the arrow goes behind the truck), to the place where you will draw an icon representing the most upstream process in the value stream.

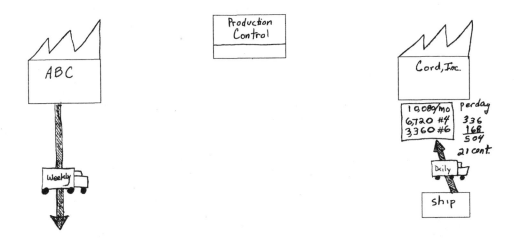

4. Draw the *manufacturing operations* along the bottom of the map, with the most upstream process on the left and the most downstream process on the right. Represent each process with an icon, label it (for example, "machining," "deburring," "testing"), and draw data boxes below each process icon. Add a notched line below the process boxes on which you will record the production timeline.

❑ Determine the number of boxes (process icons) you will need ahead of time to be sure you leave enough space for all of them.

❑ Always place the shipping icon at the far end of the process chain—directly below the truck icon for the customer.

❑ Leave enough space between each box for data. If material is stored or stands idle between processes, for example, you will need to leave space between process icons to note this and the WIP amounts (see step 7 below).

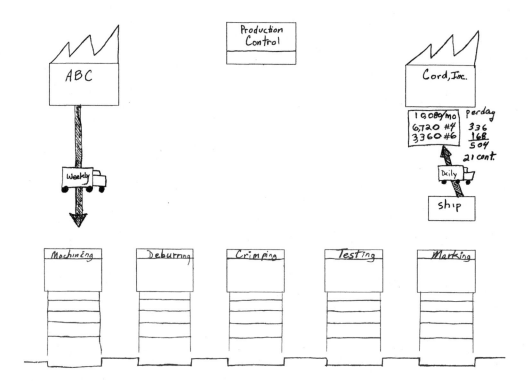

5. Enter *process attributes* in the data boxes you created under the process icons. Show the value-added time (that is the processing time or cycle time) for each operation on the line below each box. In the upper right corner of the map, enter shift time (or times), planned breaks, and total daily available production time.

Attributes Defined

- For each operation, *availability* is, in effect, the available production time. This is determined by taking the shift time or total available production time—8.5 hours (510 minutes) in the Premiere Manufacturing case study—and subtracting regular *planned* downtime occurrences (a 30-minute unpaid lunch and two 10-minute paid breaks). Therefore, the available production time is 460 minutes, or 27,600 seconds.

- Changeover times given are *per* shift.

- Changeover times are *not* considered planned downtime occurrences. Therefore, they diminish an operation's available production time to give you *actual operating time.*

- Uptime is calculated by dividing actual operating time by available time. Actual operating time for each operation equals available production time minus changeover time. For example, the uptime calculation for machining looks like this:

$$\text{Uptime} = \frac{27{,}600 \text{ seconds} - 3600 \text{ seconds}}{27{,}600}$$

Uptime = 87%

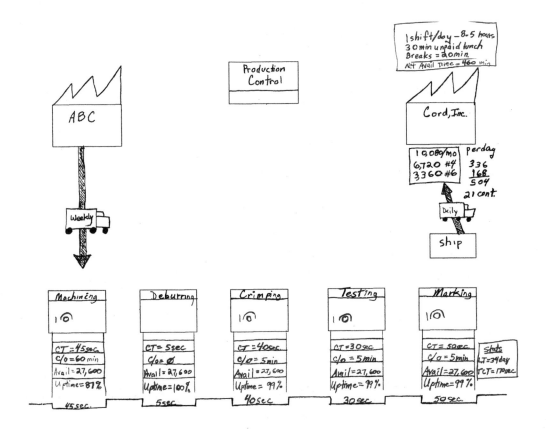

6. Show *information flow*, both electronic and manual. In most cases, information flows between customers and suppliers electronically.

❑ Draw *communications arrows* from the *customer icon* to the *production control icon*, representing the customer's forecasts and orders, and label each arrow according to frequency.

❑ Draw communications arrows from the *production control icon* to the *supplier icon*, representing production control's monthly forecast and weekly orders, and label each according to frequency.

❑ Draw a box representing the *production supervisor* in the middle of the map.

❑ Draw a communication arrow between the *production control icon* and the *production supervisor box* and label it according to the frequency with which orders are released to the supervisor.

❑ Draw communication arrows between the *production supervisor box* and the appropriate *individual process boxes*, and label each according to the frequency with which orders are released to the individual operations.

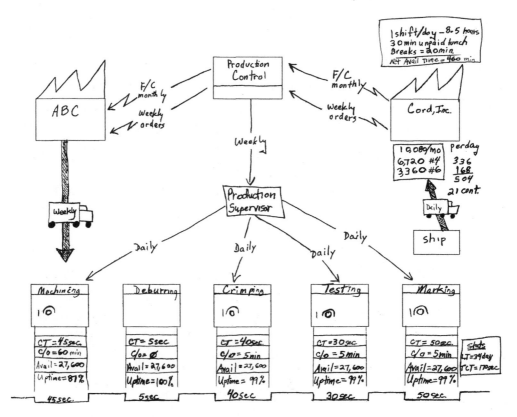

7. Draw inventory icons in the places where inventory is stored between processes. Write in the WIP quantities below the icons. Calculate the *days* of inventory on hand and write the results on the timeline that runs below the process boxes.

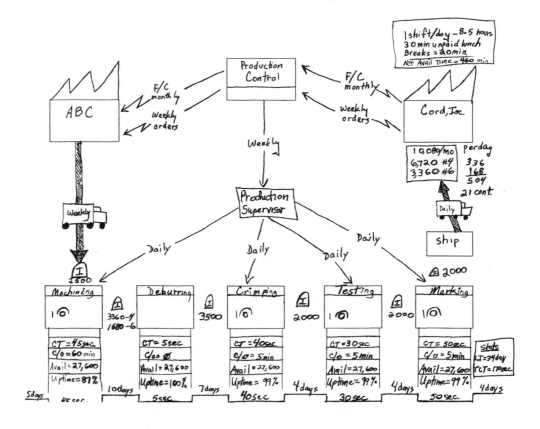

Calculating Days of WIP on Hand

You can determine days of WIP on hand by dividing the total number of hoses shipped per day (504) into the total amount of WIP between processes. For example, there are a total of 3,500 partially manufactured hoses between Deburring and Crimping. Here's how you calculate days of WIP on hand between the two processes:

3,500 hoses ÷ 504 hoses shipped per day = 6.94 or 7 days

---

8. Draw in push, pull, and FIFO locations. If a process is producing to a schedule independent of the downstream process, then you are dealing with a *push* system. Other scenarios would be some combination of pull and FIFO.

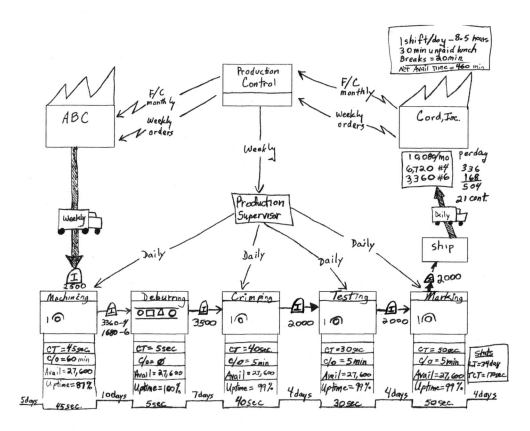

Step 4 Wrap-Up

After everyone on the core team agrees on the map's details, create a clean copy, date it, and post it on the VSM storyboard for everyone to see.

The Premiere case study that we have referenced in this chapter is a simple one, with a production line running from left to right. However, many value streams have multiple lines that merge. Regardless of the complexity of the overall process, the objective is always the same: post enough detail to show how the value stream functions, but not so much that the map is too confusing to read.

Common Problems to Avoid

The value stream mapping concept was developed by Toyota in the 1950s and attracted more widespread attention among manufacturers in 1997 with the publication of an article by Peter Hines and Nick Rich titled "The Seven Value Stream Mapping Tools."[1] According to Hines and Rich, there are several problems with the latest craze of value stream mapping:

✗ There are a range of other wastes within a value stream, such as wasted energy (e.g., lighting, heating) and the waste of human potential when human resources are under-used or their value and contribution is not recognized. Simply mapping the value stream does not capture such wastes.

✗ The benefits of mapping are extremely limited if you do not include the information flow.

✗ Typically about half of the useful information gleaned during the data collection phase of value stream mapping is subjective and informal, and thus does not show up on the map.

With these in mind, watch out for the following as you proceed with your current-state mapping.

✗ **Tunnel Vision.** Do not become so focused on the task of mapping the value stream that you are oblivious to other facts and observations. Gaining a better understanding of manufacturing processes is always a good idea, even if you cannot use what you learn in creating your value stream map. During the course of observing operations and interviewing operators, for example, you might discover that when one worker is performing a specific operation, he or she never creates defects. You would not want to neglect this important observation. Although it will not help you to create your map, it most certainly will prove to be a critical piece of information when it comes time to implement process improvements. Focusing exclusively on the mechanics of value stream mapping can also lead to the follow problems:

— inability to understand how to assign priorities to improvement opportunities.

— lack of management commitment for implementation.

— inability to tie the mapping process to a reporting mechanism for issue identification and review.

✗ **Using Value Stream Mapping Strictly as a Management Tool.** Resist any temptation to use value stream mapping strictly as a management tool. When you collect data, listening to workers' opinions and concerns is as important as measuring cycle times, changeover times, and other process attributes. Remember that becoming a lean enterprise involves engaging everyone's knowledge and creativity. If you demonstrate a sincere interest in what hourly workers have to say, you will establish trust and learn everything you want to know—and then some.

---

1 Hines, Peter, and Nick Rich. 1997 *International Journal of Operations and Production Management*, Vol 17, No. 1, pp. 46–64.

✗ **Incomplete Current-State Mapping.** Although many organizations have undertaken various steps of the Value Stream Management process, few, in our experience, tie them all together as well as Toyota. Failing to follow the process from beginning to end is a common reason for this shortcoming. Typically, companies also have difficulty modifying the technique to fit their culture; furthermore, they get preoccupied with mapping and start implementing improvements before gathering all the facts. When value stream maps fail to represent accurately a current state, they are of limited value when it comes to implementing significant, lasting improvements. On the other hand, following the Value Stream Management process from beginning to end will ensure your success. Don't forget: value stream mapping is an important tool within a structured but flexible system known as Value Stream Management! Remember that the devil is in the details. You must understand where you are currently if you hope to achieve success in implementing future state. Many future-state implementation attempts have failed precisely due to the tendency to want to rush through creating the current state. In these attempts, the tendency is to focus on the future state and not the current situation. As a result, core team members did not spend the necessary time to collect data on the current state. Do not make this mistake.

**POINTS TO REMEMBER FOR
CURRENT-STATE MAPPING**

- You must understand where you are before you can decide where you want to go.
- Focus on the most accurate and useful information.
- Gather actual information—don't use "standard" data.
- Note only the process, not the exceptions to the process.
- Don't hurry. Do it right the first time.
- Draw using icons.
- Draw in pencil or on a white board—you will make numerous changes.

---

5. Identify Lean Metrics

Step 5. Identify Lean Metrics

Now that you have documented the current state, you are ready to identify the metrics that will help you achieve your future-state goals.

The best way to get people to contribute to a lean initiative is to give them a simple way of understanding the impact of their efforts as they plan improvement activities, implement them, check the results, and make appropriate adjustments. Lean metrics provide such a tool, and help to drive continuous improvement and waste elimination (see Figure 5-1).

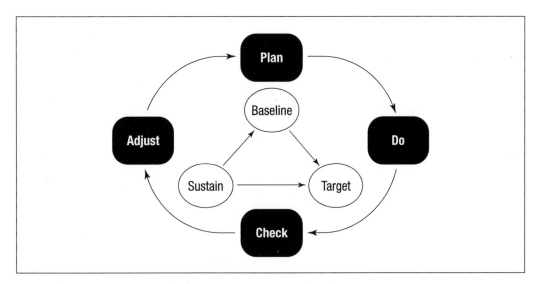

Figure 5-1. Measurement and continuous improvement cycles

As you review performance progress, be sure to highlight even small savings and improvements. These add up, and are key to the overall large gains of a lean implementation. Eliminating waste makes your company stronger and more competitive, and it results in cost reductions. We recommend that you ask your accounting staff to help you determine the financial impact of improvements such as increasing inventory turns, reducing work-in-process, reducing changeover times, and so on (see Figure 5-2).

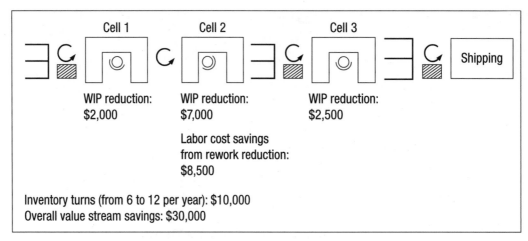

Figure 5-2. Financial impact of improvement

Lean Metrics: the Fundamentals

The metrics that best suit your organization depend a great deal on the particulars of your situation. The team charter, created during Step 1 of the Value Stream Management process, will guide you in determining the appropriate metrics. Here are some basic metrics that most companies will find useful:

→ Inventory turns.

→ Days of inventory on-hand.

→ Defective parts per million (DPPM) or sigma level.

→ Total value stream WIP.

→ Total cycle time, or total value adding time (VAT).

→ Total lead time.

→ Uptime.

→ On-time delivery.

→ Overall equipment effectiveness.

→ First-time-through capability.

→ Health and safety record (including OSHA recordable and reportable incidents).

Metrics should be easy to understand and collect. Here are some guidelines for identifying and using lean metrics:

❏ Involve those responsible for implementing change.

❏ Collect and review data when it is needed.

❏ Gather data where it is most useful.

❏ Make data accessible.

❏ Make data collection easy and reliable.

❏ Ensure that those people who can "make things happen" get timely feedback.

❏ Visually link results to specific kaizen events.

The metrics you choose should be easy to *stratify* so that they can provide specific measures for individual operations or cells as well as a total measure for the entire value stream.

Example of Metric Stratification

Uptime is the mathematical relationship between actual operating time and net available production time.

$$\text{Uptime} = \frac{\text{Actual operating time}}{\text{Available production time}}$$

To calculate actual operating time, you need to keep track of every loss that diminishes your available production time. It is possible to calculate uptime for individual operations as well as for an entire process. It is also possible—and in fact, extremely helpful—to account for the most common types of individual losses that diminish available production time:

→ Changeovers.

→ Idle time due to late delivery from an upstream operation, quality problems, material irregularities, etc.

→ Breakdowns.

The uptime metric shows at a glance how well an operation or a process uses its available time. However, stratifying the overall availability loss into the individual losses listed above provides improvement teams with a good way to prioritize improvement opportunities. For example, it's not unusual for data stratification to reveal that changeovers often account for an operation's single largest availability loss. Under such circumstances, of course, the quickest way to significant improvement is by implementing quick changeover techniques.

Steps for Identifying Lean Metrics

1. Review the list of common metrics and the specific customer targets or other improvement targets documented on the team charter. Draft an initial list of metrics.

2. Initiate a round of catchball with management to ensure their agreement with and commitment to the metrics.

3. Determine exactly how the metrics will be calculated.

4. Calculate baseline measures from the data collected during the current-state mapping process (Step 4), and post them on the storyboard.

You will determine proposed target metrics when you plan the future state, in Step 6.

PREMIERE MANUFACTURING CASE STUDY, STEP 5

After reviewing the list of common metrics and the customer's requirements, Premiere's core implementation team drafts a list of the metrics that will work best for tracking progress toward the targets. The team reviews the metrics with management and gains their buy-in. Figure 5-3 shows the metrics and the baseline figures determined during current-state mapping.

Metric	Baseline	Proposed
Total value stream WIP inventory	17,040 units	TBD
Total product cycle time	170 seconds	TBD
Total value stream lead time	34 days	TBD
On-time delivery	88 percent	TBD
Defective PPM—external	45	TBD
Uptime	84 percent	TBD

Figure 5-3. Metrics and baseline measurements

Customer requirements (for example, on-time delivery performance) heavily influenced the team members' choice of metrics, as did the observations they made and the facts they gathered while collecting data to map the current state. Doing work-in-process inventory counts between the operations was a real eye opener, and the team knows that reducing WIP will be an important improvement. The following sections show how the team calculates the baseline measurements from the current-state data.

Total Value Stream WIP Inventory

The team calculates total value stream WIP by totaling the amount of WIP inventory on-hand between each operation.

Raw material prior to machining:	2,500 hoses
Between machining and deburring:	5,040 hoses
Between deburring and crimping:	3,500 hoses
Between crimping and testing:	2,000 hoses
Between testing and marking:	2,000 hoses
Between marking and shipping (finished goods):	2,000 hoses
Total inventory:	**17,040 hoses**

The team also calculates the number of days of WIP on-hand between each operation. Daily WIP is determined by dividing the actual quantity of hoses by the daily total quantity of hoses required by the customer. The team calculates daily customer requirements (demand) by dividing the number of hoses required per month (10,080) by the number of shipping days per month (20).

Total number of hoses (#4 and #6) required per day:

10,080 hoses required per month ÷ 20 shipping days per month = 504 hoses per day

WIP calculations (#4 and #6 hoses combined):

Raw material prior to machining

2,500 hoses (504 hoses per day = 5 days on-hand

Between machining and deburring

5,040 hoses (504 hoses per day = 10 days on-hand

Between deburring and crimping

3,500 hoses (504 hoses per day = 7 days on-hand

Between crimping and testing

2,000 hoses (504 hoses per day = 4 days on-hand

Between testing and marking

2,000 hoses (504 hoses per day = 4 days on-hand

Between marking and shipping (finished goods)

2,000 hoses (504 hoses per day = 4 days on-hand

Total inventory (in days)

5 + 10 + 7 + 4 + 4 + 4 = 34 days on-hand

Total Product Cycle Time

To determine the total product cycle time, also referred to as total value adding time, the core implementation team studies the current-state map and lists the cycle times for each operation:

- Machining: 45 seconds.
- Deburring: 5 seconds.
- Crimping: 40 seconds.
- Testing: 30 seconds.
- Marking: 50 seconds.

By adding together the cycle times for each operation, the team calculates the following total product cycle time:

45 seconds + 5 seconds + 40 seconds + 30 seconds + 50 seconds = 170 seconds

PREMIERE MANUFACTURING CASE STUDY, STEP 5, *continued*

Total Value Stream Lead Time

To determine total value stream lead time, the team turns to the current-state map to see how long it takes for material to flow through the process once an order was released to the production floor. Here are the times (which, coincidentally, correspond to the number of days of WIP between the operations):[1]

- 5 days prior to machining.
- 10 days between machining and deburring.
- 7 days between deburring and crimping.
- 4 days between crimping and testing.
- 4 days between testing and marking.
- 4 days between marking and shipping.

5 + 10 +7 + 4 + 4 + 4 = 34 days

A total value stream lead time of 34 days means that it takes at *least* this long to complete a customer order. In other words, Premiere's lead time is close to seven five-day work-weeks! Given that the value-adding time or total value stream cycle time is just 170 seconds, the team realizes that the opportunity for improvement is tremendous.

On-Time Delivery

Premiere's shipment compliance or on-time delivery is only 88 percent due to the following factors:

- Uptime issues at the machining operation.
- Constantly shifting production schedules.
- 95 percent on-time delivery on average between internal suppliers and customers, resulting in cumulative on-time percentage of 81.4 percent (.95 x .95 x .95 x .95 = 81.4 percent).

Defective PPM–External

The team researches quality data and finds that the current internal defect rate, expressed in parts per million (ppm), is 5,000. The defects are mainly produced by the marking operation. Despite the high internal defect rate, Premiere passes on relatively few of the defects to Cord. However, the external defect rate, 45 ppm, is still high by lean manufacturing standards, because the goal of a lean

---

1 Although it is not always true that the lead time will equal the total number of days of WIP inventory in the value stream, the measures are often very close. This is because the time that WIP is idle and waiting to be processed is huge compared to the actual time that material is being processed.

enterprise is zero defects. The team decides that it is important to track both internal and external defect rates, but since the external defect rate is the best reflection of how well Premiere is serving its customer, this is the metric that the team chooses to display on the storyboard.

Uptime

The team determines the cumulative uptime measure for the value stream based on the uptime measures for each operation:

- Machining: 87 percent.
- Deburring: 100 percent.
- Crimping: 99 percent.
- Testing: 99 percent.
- Marking: 99 percent.

$0.87 \times 1 \times 0.99 \times 0.99 \times 0.99 = 0.844$ (84 percent)

After agreement is reached on all the data, the team transfers the information to its Value Stream Management storyboard.

Discussion Topics to Help Identify Wastes

Before proceeding to the next step—mapping the future state—it is important to identify as much waste as you can in the value stream. As you examine the current-state map from the case study, and your own current-state map, look for obvious forms of waste. Take this opportunity to review what you have learned so far with your own core implementation team, keeping track of questions as they arise. Here are some observations and questions from the case study to get you started.

1. The total value stream cycle time—or the value adding time—is 170 seconds. Yet the lead time through the value stream is 34 days! Why?

2. There is 10 days of inventory between the machining and deburring operations. Why?

3. How might you gain extra capacity without additional capital investment?

4. In traditional manufacturing environments there is often a reluctance to perform changeovers. What impact does this have on levels of WIP inventory?

5. In the Premiere Manufacturing case study, production schedules are being delivered daily to each process. Processes are independently making product, whether the downstream process requires it or not. Why are there so many production schedules?

6. What if the machining cycle time were equal to the takt time. Why would this be problematic?

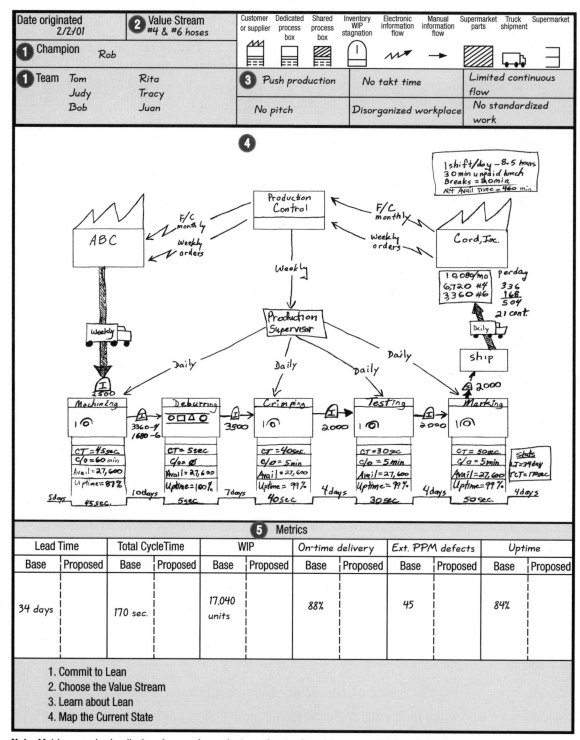

Note: Metrics can also be displayed as graphs or charts on the storyboard (see page 133 for full storyboard).

Lean Manufacturing Assessment

In the case study that appears in this chapter, you have seen a core implementation team choose some "hard metrics" against which it will track its improvement efforts.

As a complement to implementing the Value Stream Management process, many companies find it highly beneficial to complete a "Lean Manufacturing Assessment" (refer to the example on the CD-ROM) before they begin to plan and implement improvements to target value streams.

Such an assessment is often called a gap analysis. It is used to identify specific areas within the value stream on which improvement efforts can be focused, and metrics assigned. It can then be used to monitor progress over time.

The assessment begins with an effort to quantify the current level of progress related to 10 important criteria associated with lean manufacturing:

- Team involvement.
- Training.
- Workplace organization.
- Quick changeover.
- TPM.

- Quality.
- Visual controls.
- Order leveling.
- Material movement.
- Flow manufacturing.

This initial effort establishes a baseline. Then, by setting goals for each of the criteria and comparing the current rating with a goal, a "gap" is observed. Your job is to close that gap. This means, of course, that you must conduct the lean manufacturing assessment again after you have actually implemented value stream improvements to see how well you have closed the gaps that you identified initially.

Whether performed before or after value stream improvement efforts, the lean manufacturing assessment helps establish priorities (remember that improvement is a continuous process—there are always opportunities to improve). Figure 5-4 is an example of the chart that you create to display the results of your assessments.

Conducting regular lean assessments can also:

✓ Provide a standardized means of communicating broad improvement goals among various value stream projects and among different sites.

✓ Help ensure that the entire organization is on board.

✓ Help sustain and renew improvement efforts.

Here are some tips for using a lean assessment, such as the one shown on the CD-ROM:

❏ Perform the assessment as a team.

❏ Have an internal lean manufacturing expert participate to help ensure standardized application of the tool.

❏ Try to reach consensus on scoring each area, but don't get hung up on minor differences.

❏ Note the reasoning behind the scoring of each area.

❑ Don't use the lean assessment to set targets or stretch goals. It is best used as an internal "lean barometer" to track progress in a consistent way.

The lean manufacturing assessment will give you an overall idea of your performance, but it will not provide the detail you need for continuous improvement. For that you need specific lean metrics, as discussed in this chapter.

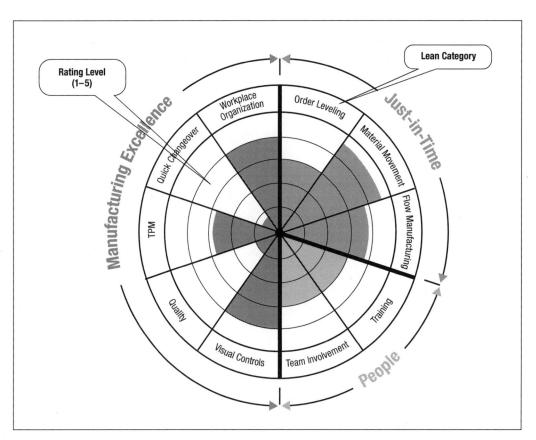

Figure 5-4. Lean assessment

6. Map the Future State

Step 6. Map the Future State

Now that you have established a picture of the current state and determined lean metrics, the next step is to tap the creativity of the workforce and the core implementation team to design the future state. Part of this process involves identifying the lean tools—such as cell design and finished-goods supermarkets—and the improvement methods—such as 5S and quick changeover—that will help ensure you can meet quality and delivery requirements. Your future-state map will show where these tools are to be used.

It bears repeating that you are still engaged in *planning* at this point. Don't get preoccupied with details. The time to be concerned about details is in steps 7 and 8.

Here we are concerned only with mapping (in effect, planning) the future state—identifying the *opportunities* to design a more efficient and waste-free value stream.

The process for mapping the future state takes place in three stages:

Customer demand stage—understanding customer demand for your products, including quality characteristics, lead time, and price.

Flow stage—implementing continuous flow manufacturing throughout your plant so that both internal and external customers receive the right product, at the right time, in the right quantity.

Leveling stage—distributing work evenly, by volume and variety, to reduce inventory and WIP and to allow smaller orders by the customer.

You learned about these three stages and the associated lean tools and concepts in Step 3. Now you will begin to apply them.

Begin the Future-State Map

Before you start mapping the future state, review the current-state map. If any questions arise regarding the current state as you plan the future state, revisit the appropriate areas on the factory floor for clarification.

We recommend that you map the future state on a flip chart using a pencil, or on a white board using dry-erase markers. Your map is bound to change as you experiment and acquire more and better information. Let your future-state map be flexible!

As we work through the future-state mapping process, we will introduce details of the Premiere Manufacturing case study to illustrate concepts.

PREMIER MANUFACTURING CASE STUDY—STEP 6, BEGIN THE FUTURE-STATE MAP

Premiere Manufacturing has committed to producing to takt time and has reached consensus with supervisors and workers. The customer, Cord, Inc., has agreed to accept daily deliveries of 504 units in reusable containers of 24 units each but would like to work toward flexible amounts. Cord will continue to send a 30-day forecast and will now begin to send orders daily. Premiere will continue to provide its supplier with a monthly forecast and weekly orders. Over the short term, Premiere has agreed to continue receiving weekly deliveries from their supplier, ABC, Inc., but ABC is willing to work toward providing more frequent deliveries if necessary.

1. Begin the future-state map by drawing the customer, supplier, and production control icons, and the communications arrows between them, in the same positions as you did on the current-state map:

❏ Draw the *customer icon* in the upper right corner.

❏ Draw the *supplier icon* in the upper left corner.

❏ Draw the *production control icon* between the customer and supplier icons.

❏ Draw communications arrows running from the customer icon to the production control icon and from the production control icon to the supplier icon.

❏ Label the arrows to show order and forecast frequencies between the customer and production control and between production control and the supplier. Be sure to make any adjustments to reflect changes in forecast or ordering frequencies you may have negotiated since you completed the current-state map.

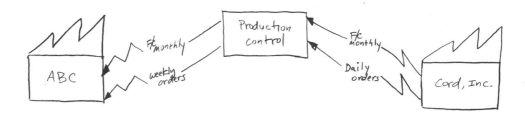

2. Now add the shipping information:

❏ Draw the shipping icon in the appropriate place.

❏ Draw a truck icon between the shipping icon and the customer icon, with a delivery arrow running from shipping to the customer. Enter the frequency with which products are shipped to the customer inside the truck icon.

❏ Draw a truck icon between the supplier icon and the icon representing the most upstream operation, with a delivery arrow running from the supplier icon to the operation. Enter the frequency with which the supplier ships raw materials inside the truck icon.

❏ Below the customer icon, enter the customer requirements (quantities).

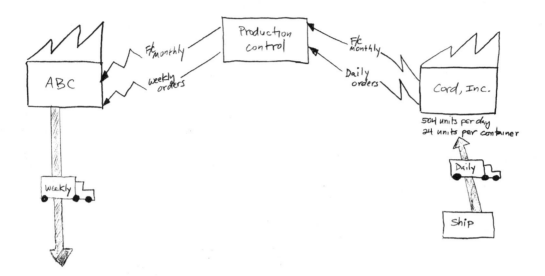

You are now ready to begin planning your future state. Here are some tips and guidelines to follow as you proceed:

❏ Do not attempt to micro-design the future state at this time.

❏ You will have to make assumptions to create targets. These assumptions may change later.

❏ Techniques planned at any stage may be modified later.

❏ Create a plan that the team can agree on.

❏ Keep a separate copy of your map at each stage, before you add elements from the next stage.

❏ This will be your first future-state map, so you may still have to make compromises. You will need to continue to improve.

Stage One: Focus on Demand

You are now ready to focus on demand. Your goal is to draw the parts of the future-state map that answer and illustrate the *customer demand* questions (see sidebar, next page).

The Customer Comes First

Undoubtedly you have heard such expressions as "the customer is number one" and "the customer comes first." Unfortunately, companies often pay lip service to customer satisfaction without actually improving it.

Lean Guidelines

DEMAND

- What is the demand? In other words, what is the takt time?
- Are you overproducing, underproducing, or meeting demand?
- Can you meet takt time (or pitch) with current production capabilities?
- Do you need buffer stock? Where? How much?
- Do you need safety stock? Where? How much?
- Will you ship finished goods right after the final operation or use a finished-goods supermarket?
- What improvement tools will you use to improve your ability to fulfill customer demand?

The good news is that by focusing on customer demand in Step 6 of the Value Stream Management process, you are doing something tangible to satisfy your customer.

Steps to Designing a Future State that Includes Customer Demand Elements

The steps shown below outline the process for designing the future-state plan to meet customer demand.

1. Determine takt time and pitch.

2. Determine whether you can meet demand using current production methods.

3. Determine whether you need buffer and safety inventories.

4. Determine whether you need a finished-goods supermarket.

5. Determine which improvement methods you will use.

Below are the new icons you will use to map the customer demand stage of the future state.

Demand Stage Icons

Purpose	Icon
Buffer Inventory	B B
Safety Inventory	S S
Supermarket	

1. Determine Takt Time and Pitch

Start the demand stage of the future state by determining your takt time. Takt time is the primary measure that determines how fast a process needs to run to match demand.

Takt time = Available production time /Required daily production quantity = Time/Volume

Note: Calculate takt time in seconds for high-volume value streams.

Next, determine your pitch. Pitch is the amount of time—based on takt—required for an upstream operation to release a predetermined pack-out quantity of WIP to a downstream operation.

Pitch = Takt Time x Pack-out Quantity

Note: Takt time is customer driven. Pack-out quantity may or may not be.

PREMIERE MANUFACTURING CASE STUDY—STEP 6, DEMAND STAGE: DETERMINING TAKT TIME AND PITCH

Takt Time

During Step 4, the team determined the available production time as 460 minutes (510 total available minutes minus 50 minutes of regularly scheduled breaks) or 27,600 seconds.

The team also determined that their customer, Cord, requires 504 hoses daily (10,080 hoses per month over a 20-day shipping month).

They calculate the takt time as follows:

Takt time = 27,600 seconds (available production time) ÷ 504 hoses (required daily
        production quantity)

Takt time = 55 seconds per hose

Pitch

Cord has requested shipments in containers that hold 24 units. The team calculates the pitch as:

Pitch = 55 seconds (takt time) x 24 units per container (pack-out quantity)
Pitch = 1,320 seconds or 22 minutes

The team adds this information to their storyboard (see following page).

At Premiere Manufacturing, a 22-minute pitch means that the manufacturing process must have 24 units available in this time instead of having one unit available every 55 seconds. We will revisit pitch when we focus on leveling production.

2. Determine Whether You Can Meet Demand Using Current Production Methods

If you are to meet customer requirements consistently, you must

❏ Determine whether you are overproducing, underproducing, or meeting demand.

❏ Determine whether you have adequate production capacity to meet demand.

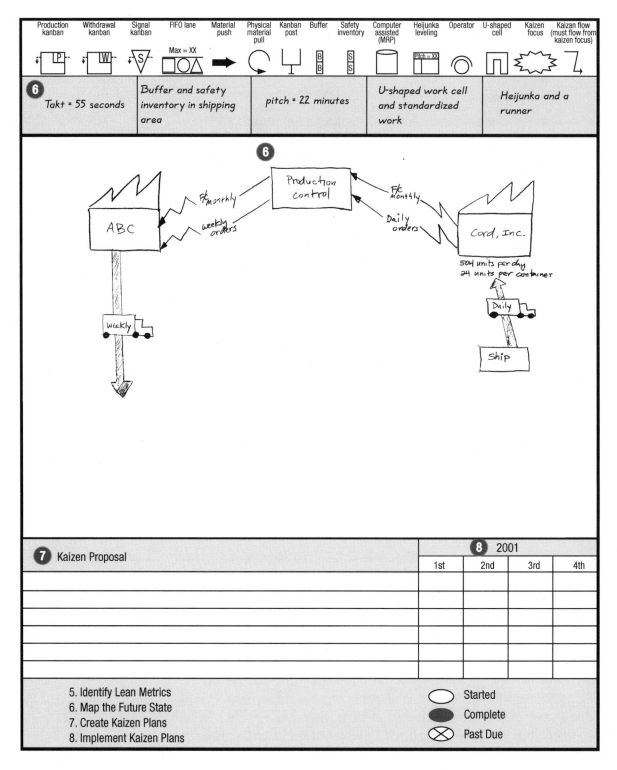

| Production kanban | Withdrawal kanban | Signal kanban | FIFO lane | Material push | Physical material pull | Kanban post | Buffer | Safety inventory | Computer assisted (MRP) | Heijunka leveling | Operator | U-shaped cell | Kaizen focus | Kaizen flow (must flow from kaizen focus) |

6 Takt = 55 seconds | Buffer and safety inventory in shipping area | pitch = 22 minutes | U-shaped work cell and standardized work | Heijunka and a runner

7 Kaizen Proposal

	8 2001			
	1st	2nd	3rd	4th

5. Identify Lean Metrics
6. Map the Future State
7. Create Kaizen Plans
8. Implement Kaizen Plans

◯ Started
⬤ Complete
⊗ Past Due

There will always be issues that prevent continuous operation, undermining your ability to meet demand consistently. To determine whether you are currently capable of meeting customer demand, review the information you gathered while mapping the current state and the baseline metrics you established in Step 5. Revisit the shop floor as needed to verify any information. You must know where the problems are so you can address them and prevent having to compensate for them by scheduling excessive amounts of overtime and shipping incomplete orders.

PREMIERE MANUFACTURING CASE STUDY—STEP 6, DEMAND STAGE: DETERMINING ABILITY TO MEET CUSTOMER DEMAND

Premiere's core implementation team learned the following facts when it established the baseline metrics in Step 5:

- On-time delivery is only 88 percent.
- Uptime is only 84 percent at a value stream level.

Premiere calculates the value stream capacity in the current state by dividing the net available production time by each of the process cycle times. They must look first at the operations with cycle times closest to takt time (55 seconds). The marking operation has a cycle time of 50 seconds and an uptime of 99%. The machining operation has the next closest cycle time of 45 seconds and the worst uptime, 87%. The team looks at these two operations to get an understanding of Premiere's ability to meet customer demand.

$$\text{Capacity} = \frac{\text{Available production time}}{\text{Cycle time for the operation}}$$

$$\textbf{Marking Capacity} = \frac{27{,}600 \text{ seconds}}{50 \text{ seconds per unit}} = 552 \text{ units}$$

Because uptime is 99%, the team adjusts process capacity accordingly:

$$552 \text{ units} \times 0.99 = 546 \text{ units}$$

$$\textbf{Machining Capacity} = \frac{27{,}600 \text{ seconds}}{45 \text{ seconds per unit}} = 613 \text{ units}$$

Because uptime is 87%, the team adjusts process capacity accordingly:

$$613 \text{ units} \times 0.87 = 533 \text{ units}$$

Given that actual capacity is close to, but still above, the required customer demand of 504 units daily, the team concludes that they could theoretically meet demand. However, their on-time delivery of 88% means they are *not* meeting customer demand.

3. Determine Whether You Need Buffer and Safety Inventories

Determine whether you need to establish buffer and safety inventories by determining whether demand is stable and whether you have the capacity and efficiencies to meet it. Remember that you are improving the value stream to ensure that you can meet customer demand *now*. You cannot afford to wait until the future state is completed. That could take six months—and very possibly, longer.

> **Buffer Inventory**—Finished goods available to meet customer demand when customer ordering patterns, or takt time, varies.
>
> **Safety Inventory**—Finished goods available to meet customer demand when internal constraints or inefficiencies disrupt process flow.

Buffer and safety inventories are a hedge against uncertainty. They create enough extra inventory to allow you to meet demand while you are implementing your kaizen plans.

Establishing buffer and safety inventories also allows you to meet demand without scrambling to schedule overtime sporadically. Instead, you can *plan* when you will run overtime—something your hourly workers will surely appreciate.

PREMIERE MANUFACTURING CASE STUDY—STEP 6, DEMAND STAGE: ESTABLISHING BUFFER AND SAFETY INVENTORIES

The team decides to establish buffer and safety inventories with one day's worth of finished goods in each. The team adds the appropriate icons to the future state map.

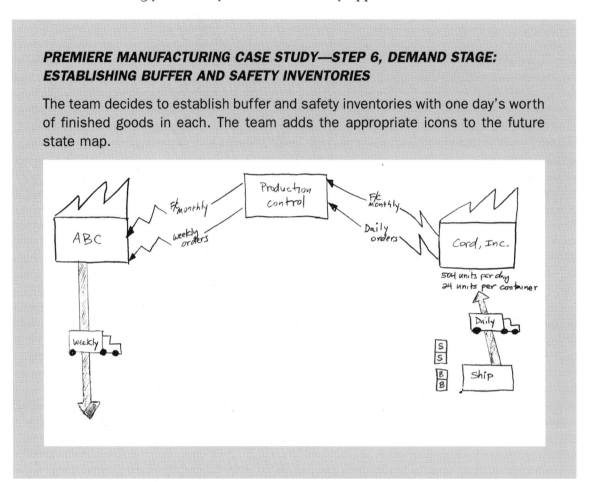

Data on overall process reliability for the value stream and the personal experience of those most familiar with individual operations will guide you in determining the appropriate levels for buffer and safety inventories. In general, neither should exceed two days' worth of finished goods.

Once you determine that buffer and safety inventories are necessary, draw the appropriate icons for them just above and to the left of the *shipping* icon.

4. Determine Whether You Need a Finished-Goods Supermarket

If you were 100 percent confident in the capability of your manufacturing process, you would ship product directly from the end of the process. But there are a variety of issues that make this difficult. Improving the value stream makes it necessary to work on reducing changeover times, line balancing, paced withdrawal, and load leveling while still ensuring that you can meet customer demand. A good means of ensuring customer demand is creating a finished-goods supermarket.

> ### *Finished-Goods Supermarket*
>
> A system used in the shipping part of the value stream to store a set level of finished goods and replenish them as they are "pulled" to fulfill customer orders. Such a system is used when it is not possible to establish pure, continuous flow.
>
> **Note:** The inventory level in the supermarket does *not* include buffer and safety inventories.

If you need to use a finished-goods supermarket, draw the *supermarket* icon between the last operation and the *shipping* icon on the future state map.

PREMIERE MANUFACTURING CASE STUDY—STEP 6, DEMAND STAGE: ESTABLISHING FINISHED-GOODS SUPERMARKET

The team decides to establish a finished-goods supermarket with one day's worth of inventory. Over the short term, it will be necessary to schedule overtime periodically to keep the supermarket stocked. The team adds the appropriate icon to the future state.

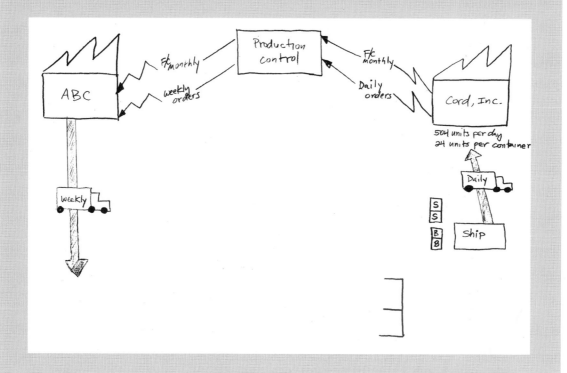

Creating a finished-goods supermarket and establishing safety and buffer inventories compromise takt image but ensure that you are able to meet customer demand. The following table summarizes the major obstacles to meeting demand and the means of dealing with them:

Obstacle	Solution
Fluctuations in customer demand	Buffer inventory
Internal problems throughout the process	Safety inventory
Inability to sustain a continuous flow from the most downstream operation to the customer	Finished-good supermarket

Remember, however, that all excess inventory is waste. Work toward minimizing or eliminating buffer and safety inventories and creating a continuous flow process in which a finished-goods supermarket is not necessary.

5. Determine Which Improvement Methods You Will Use

In keeping with the idea that it is important to work toward minimizing or eliminating "just-in-case" inventories, your future-state map should also include the methods or tools you will use to improve process capability, thus making it easier to meet demand. Some improvement methods to consider are:

→ 5S system of workplace organization and standardization.

→ Quick changeover techniques.

→ Autonomous maintenance.

→ Methods analysis and standardized work.

Although a detailed discussion of these methods and when they might be appropriate is largely beyond the scope of this book, plenty of literature that explains them in detail is readily available (see the Bibliography). Your core implementation team most likely will need to do some research to determine the most effective methods for addressing the root causes of the problems that make it difficult to satisfy customer demand.

PREMIERE MANUFACTURING CASE STUDY—STEP 6, DEMAND STAGE: DETERMINING WHICH IMPROVEMENT METHODS TO USE

The team decides that the following improvement methods will help make it possible to consistently meet customer demand:

1. **5S system.** Improving flow and eliminating obvious wastes by implementing the 5S system will pave the way for additional improvements.

2. **Quick changeover.** The 60-minute changeover at *Machining* presents a serious obstacle to producing in smaller batches. Implementing quick changeover (QCO) methods will help make the process faster and flexible enough to satisfy demand.

3. **Autonomous maintenance.** After the 5S system is implemented at each location autonomous maintenance (one of the five pillars of a TPM initiative) will be introduced at *Machining* to eliminate the small equipment problems that hurt overall process reliability.

Although it is a good idea to identify improvement methods you will use while focusing on demand, you may want to wait to draw icons for them until you have also focused on creating continuous flow. Changes recommended to create flow can alter process layout significantly. Operations may be combined or eliminated. You can add the improvement icons after you have a clearer sense of how the future state will actually look.

Stage Two: Focus on Flow

Next, you will plan and map the elements that will help you establish continuous flow. Your goal in this stage is to draw the parts of the future-state map that answer and illustrate the *flow* questions listed below.

Lean Guidelines

FLOW

- Where can you apply continuous flow?
- What level of flow do you need?
  a. "Move on, make one" (one-piece flow)?
  b. Small lots?
  c. Cell design? What type?
- How will you control upstream production?
  a. In-process supermarket?
  b. Kanban?
  c. FIFO?
- What other improvement methods will help to achieve continuous flow?
  a. Quick changeovers?
  b. Autonomous maintenance?

Continuous Flow

Implementing continuous flow—or just-in-time (JIT) production—ensures that the next downstream process has:

- Only those units needed.
- Just when they are needed.
- In the exact amount needed.

For this to happen seamlessly, the tools for customer demand—takt time, buffer and safety inventories, and a finished-goods supermarket system—must already be present so that you can work on flow without interrupting regular, timely deliveries to your customer. We also highly recommend that you implement the 5S approach to workplace standardization and organization to address many of the physical and psychological barriers to change.

Steps to Designing a Future State that Includes Continuous Flow Elements

The steps shown below outline the process for designing continuous flow into a manufacturing process.

   1. Perform line balancing.
   2. Plan for work cells.
   3. Determine how to control production upstream.
   4. Determine which improvement methods you will use.

In addition to the icons you have used to map the current state and the demand stage of the future state, you will need some or all of the icons shown below to complete the flow stage of your future-state map.

Flow Stage Icons

Purpose	Icon
Kanban Post	
Production Kanban	
Withdrawal Kanban	
Signal Kanban	
Supermarket Parts	
U-shaped Cell	
Computer-assisted (MRP)	
Physical Material Pull	

1. Perform Line Balancing

The first step in creating continuous flow is line balancing. This will help you optimize the use of personnel and balance the workload to achieve a smoother flow.

> **_Line Balancing_**
>
> Line balancing is the process by which you evenly distribute the work elements within a value stream in order to meet takt time.

To balance a line you need to:

❏ Review the current cycle times and work element assignments.
❏ Create an operator balance chart.
❏ Determine the number of operators needed.
❏ Plan the changes needed to balance the work among the target number of operators.
❏ Create a future-state operator balance chart.

PREMIERE MANUFACTURING CASE STUDY—STEP 6, FLOW STAGE: LINE BALANCING

The core implementation team decides to redesign the line for better flow. The first thing team members do is review the current-state data, as shown in the table below.

	Machine	Deburr	Crimp	Test	Mark
Cycle time	45 sec.	5 sec.	40 sec.	30 sec.	50 sec.
Changeover	60 min.	0	5 min.	5 min.	5 min.
Operators	1	0	1	1	1
Uptime	87%	100%	99%	99%	99%
Availability	27,600 sec.	27,600 sec.	27,600 sec.	27,600 sec.	27,600 sec.

Next, the team creates an operator balance chart to compare the cycle times of each operation to takt time. The balance chart shows that the line is clearly out of balance (Figure 6-1).

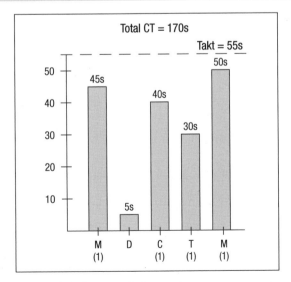

Figure 6-1. Operator balance chart—current state

PREMIERE MANUFACTURING CASE STUDY—STEP 6, FLOW STAGE:
LINE BALANCING, continued

By dividing the total cycle time of 170 seconds by the 55-second takt time, the team calculates that it should be possible to achieve takt time with three operators (currently there are four).

Formula for Determining the Number of Operators

$$\text{Number of operators} = \frac{\text{total cycle time}}{\text{takt time}}$$

$$\textbf{Number of operators} = \frac{\textbf{170 (total cycle time)}}{\textbf{55 (takt time)}} = \begin{array}{l}\textbf{3.09 operators}\\ \textbf{(or 3 operators)}\end{array}$$

After reviewing the current state, going to the shop floor, and talking to operators and the production supervisor, the team decides to set a total cycle time target of 150 seconds. Achieving this total cycle time should require only 2.7 operators.

$$\textbf{Number of operators} = \frac{\textbf{150 (total cycle time)}}{\textbf{55 (takt time)}} = \textbf{2.7 operators}$$

Based on its discussions with operators, the team believes it can achieve the new target by making the following changes:

❑ Improving programming and tooling maintenance to eliminate the need for deburring—one operator required.

❑ Distributing the work of the three remaining operations between two operators.

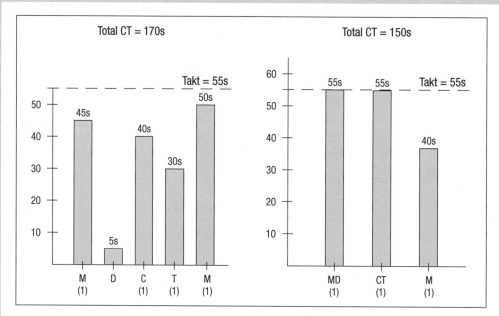

Figure 6-2. Operator balance chart—current and proposed

❑ Reducing changeover times in machining to 15 minutes each. (Assuming two changeovers per shift, this means a total of 30 minutes of changeover time per shift.)

❑ Achieving changeover times of less than one minute in the crimping, testing, and marking operations.

The team decides that the value stream should operate with three operators. Although the personnel requirement was calculated as 2.7 operators at a total cycle time of 150 seconds, there is simply too much for just two operators. The operator balance chart for the future state is shown in Figure 6-2.

2: Plan for Work Cells

The use of work cells promotes one-piece flow, because in work cells equipment and personnel are arranged in process sequence and the cell includes all the operations necessary to complete a product or a major production sequence. At this point in your future state planning, review the operations in the line for likely operations to redesign into a work cell arrangement.

PREMIERE MANUFACTURING CASE STUDY—STEP 6, FLOW STAGE: CELL DESIGN

The team realizes that achieving a balanced line depends on applying the principles of cell design. The plan is to keep machining as a stand-alone operation and combine crimping, testing, and marking into one cell. Team members add icons representing the following items to the future-state demand map:

1. Two future-state cells:
   • Machining.
   • Crimping/testing/marking.

2. New attributes for each cell:
   • Machining—
     – total cycle time: 45 seconds
     – changeover: 15 minutes
     – # of changeovers: 2 per shift
     – uptime: 93 percent
   • Crimping/testing/marking—
     – total cycle time: 105 seconds
     – takt time: 55 seconds
     – changeover: less than 1 minute

The team decides that no major redesign is necessary in machining because 5S implementation will address any equipment problems that might arise.

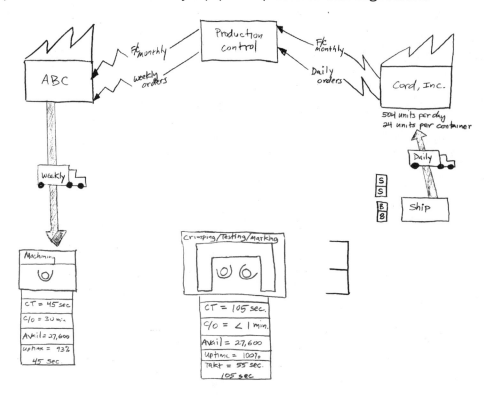

3. Determine How to Control Upstream Production

At points in the system where continuous flow is not achievable, you must determine how you will control the flow of production. Review the process for places where obstacles to continuous flow exist, and determine whether you can:

- Use in-process supermarkets.
- Use a kanban system.
- Use FIFO lanes.
- Use computer-assisted scheduling (MRP).

PREMIERE MANUFACTURING CASE STUDY—STEP 6, FLOW STAGE: CONTROLLING UPSTREAM PRODUCTION

Team members are confident that reconfiguring the old line into a process with two cells is a good way to sustain flow. They believe that pulling this off will require in-process supermarkets prior to machining and between machining and crimping/testing/marking. A kanban system will also be critical to sustaining flow. The team decides to defer mapping this until it focuses on leveling production, when it will have a better handle on how material will actually flow through the value stream to meet customer demand.

In the meantime, the team draws two in-process supermarkets on the future state map in the positions mentioned above.

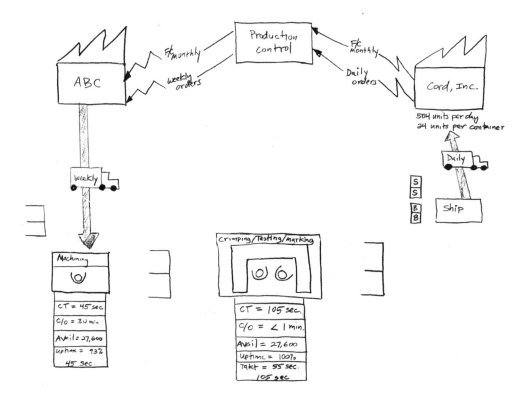

4. Determine Which Improvement Methods You Will Use

As you plan to work toward the goal of achieving one-piece flow, your future-state map should include the methods or tools you will use to improve flow. Some improvement methods to consider again at this stage include:

→ 5S system.

→ Quick changeover.

→ TPM.

→ Autonomous maintenance.

→ Standardized work.

PREMIERE MANUFACTURING CASE STUDY—STEP 6, FLOW STAGE: DETERMINING WHICH IMPROVEMENT METHODS TO USE

The team has already identified some improvements necessary for achieving targets. Now it reconsiders its list of improvement methods and draws icons at the appropriate places on the future-state map. Here is the list of the improvement methods the team believes will be necessary for creating and sustaining continuous flow:

1. 5S, TPM, autonomous maintenance, and QCO at machining.

2. Standardized work at the crimping/testing/marking cell.

PREMIERE MANUFACTURING CASE STUDY—STEP 6, FLOW STAGE: DETERMINING WHICH IMPROVEMENT METHODS TO USE, *continued*

3. QCO at the crimping/testing/marking cell.

4. 5S at the crimping/testing/marking cell (to address problems with new layout) and at shipping.

By identifying TPM as an improvement method, the core implementation team in the Premiere Manufacturing case study is not calling for a true, plantwide TPM initiative, but an isolated application of TPM methods.

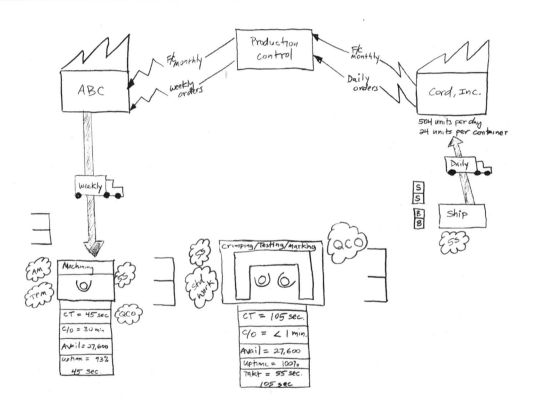

Stage Three: Focus on Leveling

In the final stage of your future-state mapping, you will continue by adding to your map the elements that will help you level production. Your goal in this section is to draw the parts of the future-state map that answer and illustrate the leveling questions (see sidebar, next page).

Leveling Production

Leveling production means evenly distributing over a shift or a day the work required to fulfill customer demand. If you do not level production some cells will fall behind in production (causing idle time downstream) while at other times they may be waiting for work.

Lean Guidelines

LEVELING

- What types of kanban cards will you use?

- How will kanban cards be distributed?

- Where in the process will you schedule production requirements?

- Will you use a heijunka box?

- What will the runner's route be?

When we focused on demand, we talked about the importance of takt image—the vision of an ideal state in which you have eliminated waste and improved the performance of the value stream to the point that you have achieved one-piece flow based on takt time. Circumstances may make it difficult to achieve true one-piece flow, but you must still strive to maintain takt image, matching the pace of your production system to the pace of sales or takt time.

Leveling production is the means to achieving such a goal.

How to Level Production

Leveling production means designing a system in which information flow regarding customer demand is smoothly integrated with the flow of material through the value stream. To make this happen you must base your flow either on paced withdrawal or a heijunka system.

The steps shown below outline the process for designing a leveled manufacturing process.

1. Decide on the best method for monitoring production against the pace of sales—paced withdrawal or a heijunka system, and design or refine the kanban system, if necessary.

2. Determine the route of the runner (material handler) and map all material and information flow.

3. Determine which improvement methods you will use, and add useful data to the map.

In addition to the icons you have used to map the current state and the demand and flow stages of the future state, you will need the icons shown below to complete your future-state map.

Purpose	Icon
Heijunka Box	Pitch = XX
Runner Route	

1. Decide on Paced Withdrawal or a Heijunka System and Design or Refine the Kanban System

Paced withdrawal can be used to move small batches when there is no variety in the value stream and pitch increments are identical. A heijunka system will help you level production based on the volume and *variety* of product being manufactured. Review your current state data and the demand and flow stage future-state maps, and determine whether a heijunka system will be required. If you have decided on using a kanban system to control upstream production, you should also review and plan for the kanban system at this point.

PREMIERE MANUFACTURING CASE STUDY—STEP 6, LEVELING STAGE: DECIDE ON PACED WITHDRAWAL OR A HEIJUNKA SYSTEM

The team reviews the following three facts, most of which were determined earlier in the process of creating the future state:

- It will be necessary to create a kanban system
- The customer has requested a container size of 24 units per container.
- The containers can be reusable

Pitch has not changed since it was calculated while focusing on demand, because takt time (55 seconds) and container size of 24 (pack-out quantity) have not changed.

Pitch = 55 (Takt Time) × 24 (Pack-out Quantity) = 1,320 seconds (22 minutes)

This means that every 22 minutes a container of 24 units must be packed and ready to ship.

Kanbans

The team decides that the value stream requires the following types of kanbans in the following locations:

1. Withdrawal kanbans that will tell the runner (material handler) how many units should be pulled from the finished-goods supermarket and staged in *shipping*.
2. Production kanbans that tell operators in the crimping/testing/marking cell how many units must be produced to replenish those pulled from the finished-goods supermarket.
3. Signal kanbans at the in-process supermarket between machining and the crimping/testing/marking cell that tell the machining operator how many units have been pulled from the supermarket.
4. Signal kanbans just upstream of machining that tell the supplier how many units have been pulled from raw material inventory.

The team decides that machining will produce in batches of 96 to 192 units. The batch size, a multiple of the 24-unit pack-out quantity, will depend on the number of signal kanbans that have accumulated after machined parts have been pulled from the supermarket between *Machining* and the *Crimping/Testing/Marking* cell.

The team decides to wait to draw the icons representing the kanbans. It will add the signal kanban icons when it maps material and information flow in *Machining*; it will add the *withdrawal kanban* and *production kanban* icons when it maps the runner's route.

Heijunka or Paced Withdrawal?

The team decides to implement heijunka for the following reasons:

1. A heijunka box, a device used to implement a heijunka system, provides a great means of visually controlling withdrawal kanbans.

2. The team talks about the likelihood of creating one value stream to produce #4, #6, #8, and #10 hoses, because the processes are so similar. Given this likelihood, the team reasons that it would be wise to familiarize people with heijunka now so that it does not seem like such a big change when the value stream begins to produce four hoses instead of two.

The team draws a *heijunka box* icon below the *production control* icon, with a manual communication arrow running from production control to the heijunka box. The entire pull system begins here. Since the ratio of customer demand for the #4 and #6 hoses is 2:1, production control will place kanban cards in the heijunka box in such a way that Premiere makes two lots of #4 hoses for every lot of #6 hoses that it produces (Figure 6-3).

Value stream: −4 and −6 collant hoses
Pitch = 22 minutes

	6:00 – 6:22	6:22 – 6:44	6:44 – 7:06	7:06 – 7:28	7:28 – 7:50	Break – NP	8:00 – 8:22	8:22 – 8:44	8:44 – 9:06	9:28 – 9:50	9:50 – 10:12	Lunch – NP	10:42 – 11:04
−4	1	1		1	1			1	1		1		1
−6			1				1			1			1

Figure 6-3. Representation of heijunka box, Premiere Manufacturing Case Study

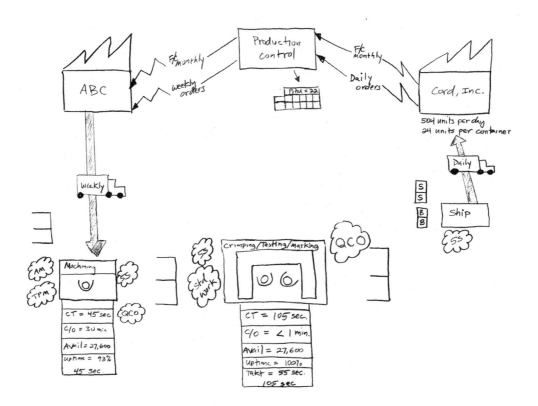

2. Determine the Route of the Runner (Material Handler) and Map All Material and Information Flow

The runner, or material handler, will ensure that pitch is maintained throughout the process. The runner will follow a designated route, timed to work within the pitch period, picking up and delivering kanban cards, components, and tooling as needed. As you are planning the functioning of your future state, now is the time to visualize the route of the runner, who will keep things moving. It is also necessary to plan how the material and information flow will work throughout the system where pull will be established, where push remains a necessity, and how communications between operations will function.

3. Determine Which Improvement Methods You Will Use and Add Useful Data

As with the demand and flow stages, your leveling stage future-state map should include the methods or tools you will use to achieve your goals. Methods to consider include:

→ 5S system.

→ Visual controls.

→ Improvement Methods.

Complete the future-state map by adding a *statistics* box showing value stream lead time and total cycle time. Show days between operations and/or cells, and add any other useful data to the map (see storyboard on page *133*).

PREMIERE MANUFACTURING CASE STUDY—STEP 6, LEVELING STAGE: DETERMINE THE ROUTE OF THE RUNNER, MAP MATERIAL AND INFORMATION FLOW, AND IMPROVEMENT METHODS

Runner's Route, Step 1

Now the team begins to draw the runner's route:

1. A stick figure drawn beneath the *heijunka box* icon represents the runner.

2. A team member draws a *withdrawal kanban* icon halfway between the icon representing the runner and the icon representing the finished-goods supermarket. A dashed arrow running from the *runner* icon to the *supermarket* icon is added. The arrow bisects the *withdrawal kanban* icon.

Runner's Route, Step 2

1. The team draws a dashed arrow from the *finished-goods supermarket* icon to the *Shipping* icon to represent the next leg of the runner's route.

2. Just beneath the dashed arrow, the team draws a *supermarket parts* icon and a *physical material pull* icon to show that the runner pulls parts from the supermarket to stage for shipment.

Runner's Route, Step 3

1. The team draws a *production kanban* icon just above and slightly to the right of the icon representing the *Crimping/Testing/Marking* cell.

2. The team draws a dashed arrow running from the *Shipping* icon to the *Crimping/Testing/Marking* cell icon. The arrow bisects the *production kanban* icon.

Runner's Route, Step 4

The team draws a dashed arrow from the *Crimping/Testing/Marking* icon to the *finished-goods supermarket* icon. This is all that is required to show that the runner transfers finished goods from the *Crimping/Testing/Marking* cell to the supermarket.

Runner's Route, Step 5

After the runner transfers finished goods to the finished-goods supermarket, he returns to the heijunka box to retrieve the withdrawal kanban from the next slot. The team draws a dashed arrow from the *finished-goods supermarket* icon to the *heijunka box* icon to document this last leg of the runner's route.

You may wonder where the runner gets the production kanbans. Production kanbans are attached to containers of finished goods. When a runner pulls a container of finished goods from the super- market to stage for shipment, he or she also pulls the production kanban from the container. The production kanban is then used to initiate replenishment of the inventory that has just been pulled.

Mapping Material and Information Flow in Machining

1. An operator from the *Crimping/Testing/Marking* cell will be responsible for pulling required parts from the supermarket between that cell and *Machining*. This is shown on the map by drawing a *supermarket parts* icon and a *manual material pull* icon between the supermarket icon and the *Crimping/Marking/Testing* cell icon.

2. When the *Crimping/Testing/Marking* cell operator pulls machined parts from the in-process supermarket, the operator will also pull a signal kan- ban from the container and place it in a special holder on the side of the supermarket flow rack. The *Machining* operator will retrieve signal kanbans when he or she delivers machined parts to the supermarket flow rack. To illustrate this part of the plan the team draws a manual communication arrow and a *signal kanban* icon from *supermarket* icon to the *Machining* icon. It draws a material push arrow running from the Machining icon to the in-process supermarket icon before *Crimping/Testing/Marking*.

Note: Even though *Machining* is not overproducing parts, a material push arrow is still drawn between the operation and the in-process supermarket just down- stream. This indicates that the production schedule in *Machining* is not based strictly on what the *Crimping/Marking/Testing* cell has just pulled from the in- process supermarket.

3. The *Machining* operator will be responsible for pulling containers from the raw materials supermarket just upstream. To illustrate this plan, the team draws a *manual material pull* icon and a *supermarket parts* icon between the *supermarket* icon and the *Machining* icon.

4. The *Machining* operator will also be responsible for pulling signal kanbans from raw material containers and placing the signal kanbans on a special kanban post. The supplier's truck driver will be responsible for collecting the signal kanbans and taking them back to the supplier's plant. To illustrate this part of the plan, the team draws a *kanban post* icon between the supermarket icon and the *supplier truck* icon. A manual communication arrow and a signal kanban icon are drawn running from the kanban post icon to the supplier icon.

Improvement Methods

The team decides to use Visual Controls, or Visual Workplace (VW), at the crimping/testing/marking cell and the heijunka box. It adds these icons to the map, along with lead times and total cycle time.

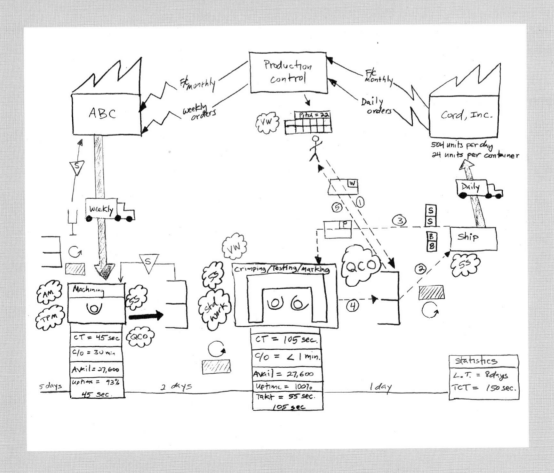

At this point, you have followed along in the Premiere Manufacturing case study and seen how a future-state map based on an understanding of customer demand, continuous flow, and leveling is created. This is the map that will be posted on the storyboard.

Why Maintaining Takt Image Matters

Making the effort to understand and fulfill customer demand, to promote flow, and to level production is not enough to ensure a successful lean transformation. Your success depends a great deal on your ability to identify and resolve problems quickly. In other words, a fast and flexible manufacturing system is also one that detects and corrects defects, errors, and other variances as quickly as possible. This is why maintaining takt image is so important.

The following question may have occurred to you at some point as you read the Premiere Manufacturing case study: Why would you move product through the value stream at 22-minute intervals when the customer only requires one shipment per day? Wouldn't it be more practical to move product from the end of shipping perhaps every four hours or every two hours?

Maintaining a takt image by using a runner who withdraws product from the finished-goods supermarket every 22 minutes sends a message about meeting demand—either you are or you aren't. If you release customer orders to shipping every four or eight hours you simply can't respond to problems that arise as quickly as you can when orders are released every 22 minutes.

Step 6 Wrap-Up

- ❏ Review the future-state plan with anyone else who may be able to suggest additional improvements.
- ❏ Modify the plan if necessary before going on to Step 7.
- ❏ For each baseline metric determined in Step 5, determine the corresponding proposed measure for the future state.
- ❏ Be sure to recognize the core implementation team's efforts to date.
- ❏ Post the most current version of the future-state map on the appropriate place on the storyboard.

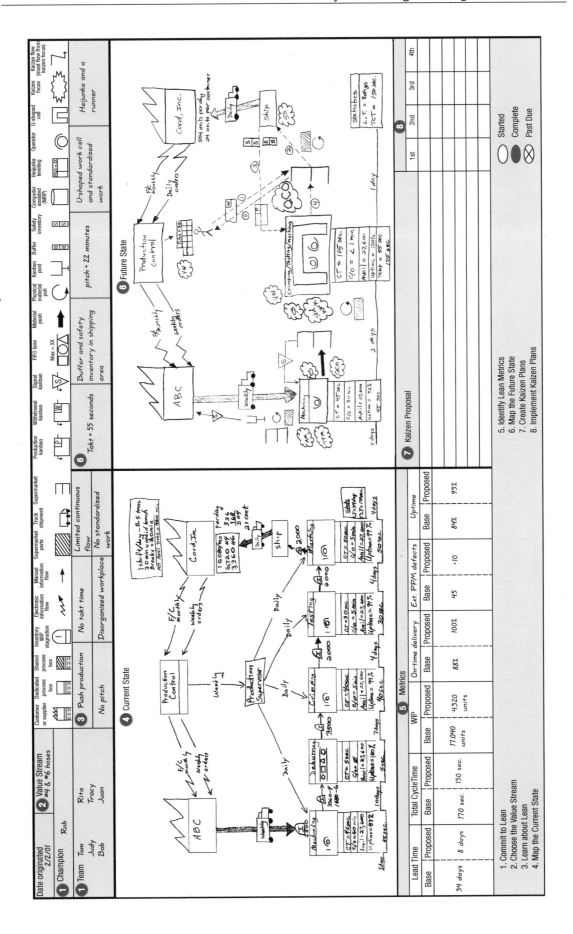

7. Create Kaizen Plans

Step 7. Create Kaizen Plans

By the time the core implementation team gets to Step 7, they may be chomping at the bit to begin implementing the future state. However, now is the time for the team to create detailed plans that will guide efforts to improve the value stream. Without solid planning, the chances for a successful lean transformation are slim.

In the planning stage, schedule recurring meetings to ensure that communication flows and that all involved parties remain on the same page. In partnership with your value stream champion, utilize the Value Stream Project Status form as milestones are attained. (An example is provided below. This is also a good tool for highlighting concerns and their potential solutions.)

Value Stream Project Status: June 30
Schedule:
• Target completion date of December 30th
Accomplishments:
• Drafted Charter
• Conducted training class on VSM
• Modified product family to #4 and #6 only
Concerns (Issues):
• Ability to maintain meetings at 3:00 p.m.
• Utilize a more comprehensive planning tool to schedule improvements
Plan:
• Schedule meetings for 1:00 p.m.
• Team Leader to utilize Kaizen Milestone Worksheet for the detailed implementation of Demand, Flow and Leveling Focused Kaizens

Comprehensive planning is very important, but keep in mind that your plans will not be perfect. You will, in fact, find yourself modifying plans as you proceed through implementation and gain more practical experience with lean manufacturing methods.

Value Stream "Kaizen" Stages

To increase your chances for success, plan your implementation using the same three stages you used to plan the future state:

- Plan how you will ensure that you are able to meet *customer demand*.
- Plan how you will improve the *process flow*.
- Plan how you will *level production*.

Using this planning sequence allows for the most effective and least costly implementation of kaizen plans. For example, once you truly understand customer demand, you may gain additional insight into how to arrange operations into cells. Similarly, the nature of your leveling system may change once you have some real experience with continuous flow production.

As you proceed with the kaizen planning process, it will help to follow these steps:

1. Review the future-state map and create a monthly kaizen plan.
2. Determine milestones for each major kaizen activity and create a kaizen milestone chart.
3. Complete the Value Stream Management storyboard.
4. Obtain management approval for kaizen plans through "catchball."

The Monthly Kaizen Plan

Start by creating a detailed monthly kaizen plan plotting the implementation schedule for the main elements or "events" to be accomplished in each stage of your future-state map (demand, flow, and leveling). Use symbols to plot the timeframes for each main event:

- ❑ Open triangles signify start dates.
- ❑ Dashed lines signify the expected duration of implementation.
- ❑ Closed triangles signify completion dates.

The monthly kaizen plan provides a macro-level sequence or structure for your implementation. (A partial example of a monthly kaizen plan is shown on the next page.)

Milestones: Break the Plan into Manageable Pieces

With the overall structure for the main improvement events in place, you can now create a series of specific milestones for each event. Milestones are definable activities or tasks that are required to accomplish the improvement, and that have been assigned completion dates. Examples include:

- Create a staging area in shipping.
- Install gravity racks at all supermarket locations.
- Establish production based on pitch.

Step 7: Monthly Kaizen Plan Worksheet

Value Stream: _____ Date: _____

STAGE	Specific Event	Six-Month Schedule					
		J	F	M	A	M	J
D							
E							
M							
A							
N							
D							

Create a kaizen milestone chart to document and track the completion of milestones. Kaizen milestone charts are similar to monthly plans in that they show the sequence of implementation activities and their completion over time. However, kaizen milestone charts get down to the next level of detail. Monitor your milestone activities on an ongoing basis to track progress toward completion. Use the same symbols as you used in the monthly kaizen plan.

A kaizen milestone chart from the Premiere Manufacturing case study is shown in Figure 7-1.

In this example, the first callout shows a solid triangle in the second week of June. This signifies that a staging area was established in shipping in the time allotted during Step 7. As you know, things do not always go according to plan. Such is the situation highlighted by the second callout. Here, the installation of gravity racks at all supermarket locations was scheduled to have started the second week of June and to have been completed two weeks later. That the dashed line extends two weeks into July means that the milestone was achieved two weeks later than planned.

Sometimes, however, milestones are achieved earlier than planned. Such is the situation described by the third callout. The core implementation team allotted five weeks to establish production based on pitch. In actuality, this milestone was achieved in three weeks.

Complete the Value Stream Management Storyboard

Whether you have been building your storyboard step by step or waiting to post information to it, now is the time to complete it. Add your proposed monthly kaizen plan and any other data you may have overlooked earlier. The completed version of the storyboard for the Premiere Manufacturing case study is printed on the inside back cover of this book.

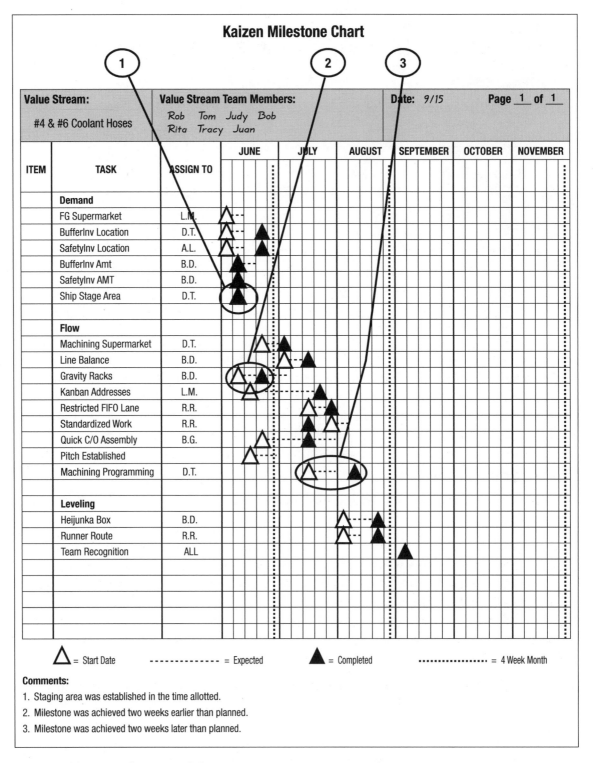

Figure 7-1. Kaizen milestone worksheet

Catchball

After you complete your storyboard, it is time to play catchball and get buy-in for your plans. All your hard work—particularly the effort to generate kaizen plans—really pays off at this point. The storyboard summarizes your plans for a lean transformation of the target value stream. You will most likely be presenting your plan to high-level managers. If you show that you have planned the future state carefully and given careful thought to *how* the plan will be implemented systematically, the catchball process will go smoothly.

 A good way for the core implementation team members to prepare for the meeting at which they present the storyboard to upper management is for the team to review and be sure they can answer the following questions:

- Why are we implementing lean in this value stream?
- What impact will implementing lean methods have on our customers?
- What quality improvements will we achieve?
- What cost savings will we achieve?
- What lead-time reductions can we achieve?
- How does this project relate to our strategic objectives?

Being able to answer questions like these will demonstrate that the team has a solid understanding of the strategic intent behind the lean transformation, and will help the team gain approval for the plan.

Planning Recap

Remember that planning is primarily about managing action. Here are some guidelines for effective kaizen planning:

- ❏ **Be realistic**—especially regarding completion dates.
- ❏ **Play catchball**—get buy-in from all stakeholders.
- ❏ **Be detailed**—detailed enough to promote clear communication and understanding.
- ❏ **Communicate**—show proposals to everyone connected to the value stream.
- ❏ **Make it visual**—use the storyboard.
- ❏ **Recognize good work**—make sure people's contributions are recognized (not only by management, but also by peers).
- ❏ **Be sure to celebrate**—you've done a tremendous amount of work. You deserve to celebrate. Celebration is an important way to recognize people's contributions.

Prepare for Implementation

If you have planned the future state and its systematic implementation carefully, it is unlikely you will need to make anything other than minor adjustments to your plans. Make whatever modifications are required, then prepare for implementation.

It is likely that various members of the core implementation team will be leading a series of kaizen events to make the changes necessary to transform the value stream. A kaizen event is a team event dedicated to quick implementation of a lean manufacturing method in a particular area over a short time period.

Be prepared for lots of hard work—but also fun. Under the umbrella of the Value Stream Management process, kaizen events and value stream mapping are tremendously more meaningful and productive than they were when conducted as isolated events.

Much has been written on the subject of preparing for kaizen events, so we will just emphasize the main points:

❑ Identify clearly the objectives of the event and be sure to communicate them clearly to the kaizen team before team members go out onto the floor.

❑ Identify any training that needs to take place during the event.

❑ Define the scope of the kaizen team's efforts.

❑ Register the team by submitting a team charter (if required).

❑ Use the kaizen milestone chart to determine the projected completion date for events, but be flexible.

❑ Determine special needs that need to be resolved to prevent problems from occurring.

❑ Draft an agenda for the duration of the event.

Value Stream: #4 and #6 hose line
Value Stream Stage: Demand
Event Date: May 17–18

Kaizen Event Objective:
Establish a finished-goods supermarket system

Day 1
- Welcome and Introduction
- Review VSM Storyboard—Get everyone on the same page!
- Adjust any parts of the demand future state
- Identify issues
- Walk the actual flow as defined in the future state
- Conduct training on supermarkets and kanbans
- Begin analysis of kanban flow and conduct kanban calculation

Day 2
- Begin setting up stores
- Create kanban cards
- Set up address locations for the supermarket
- Determine list and color code system for stores
- Trial kanban system
- Collect data and monitor
- Review data
- Write initial supermarket and kanban procedures
- Update job descriptions
- Update internal procedures
- Monitor
- Set plan to standardize within 1 week
- Team recognition

Figure 7-2. Implementation agenda

Shown below is a detailed implementation agenda for a kaizen event from the Premiere Manufacturing case study storyboard. Note that the last item on the implementation agenda shown is "team recognition." We can't emphasize enough how important it is to recognize people's efforts as they go through the stages of implementing lean. Transforming a value stream to its lean future state may take a year—or longer. If people work long periods without feeling appreciated, they are bound to lose focus and enthusiasm.

7 Kaizen Proposal	**8** 2001			
	1st	2nd	3rd	4th
Demand—Establish buffer and safety stock, FG supermarket, implement 5S	◯	●		
Flow—Cell design, line balancing, and standardized work	◯	●		
Flow—Reduce changeover times, implement autonomous maintenance		◯	●	
Level—Establish runner's route		◯		
Level—Implement heijunka and kanban system		◯		●
Level—Additional TPM applications: modify PM standards				◯

5. Identify Lean Metrics	◯ Started
6. Map the Future State	● Complete
7. Create Kaizen Plans	⊗ Past Due
8. Implement Kaizen Plans	

8. Implement Kaizen Plans

Step 8. Implement Kaizen Plans

All the planning and preparation done so far should now allow you to proceed to the implementation phase with enthusiasm and confidence. However, remember that when implementation begins in earnest, kaizen activities will have an impact on virtually everyone connected to the target value stream.

Recommendations for Coping with Change

Change—even change for the better—is difficult for most people. But the more people know about what's going on, the easier it is to deal with the anxieties that accompany significant change. Below are some recommendations for managing and coping with the anxieties that are bound to present themselves as you proceed.

❏ **Communicate, communicate, communicate!** Make sure that everyone upstream and downstream of the area where a kaizen event is taking place knows what is happening and why. A brief explanation by a team leader or a supervisor at a shift-start meeting may be all that is necessary to assure people that no one is withholding information from them.

❏ **Address negative behavior early in the implementation.** If someone does not seem to be participating, or displays negative behavior, talk to that person privately. Listen to his or her concerns and work to resolve them. Active listening is an acquired skill. Hear people out; show that you genuinely care. *Then* respond. Explain how the improvement effort will make the company stronger, which will potentially make everyone's future brighter and more prosperous. If possible, assure people that no one will lose his or her job as a result of improving the value stream.

❏ **Do not let a problem stop the process.** Perhaps an unforeseen problem makes it impossible to complete a kaizen event. Acknowledge the problem and reschedule the event for completion as soon as the problem is resolved. Do not look at the delay as a failure, but as a detour in a fascinating journey!

❏ **Consider each kaizen event an experiment.** Let's say that a team is doing cell design, but you underestimated how much time it would take and didn't build up enough safety inventory. Now you really have to scramble so that you don't shut down your customer's assembly line. Perhaps next time you should do cell design over the weekend. You *will* make some mistakes. Learn from them and move on!

❏ **Reward and recognize people's efforts, practice mutual trust and respect, and treat people with honesty and integrity every day.**

❑ **Be present.** The value stream champion and top managers should go to the floor regularly to encourage people and to find out what they can do to support improvement efforts.

❑ **Be flexible.** Unexpected things will happen, but flexibility combined with focus and commitment will win the day.

Keep the Big Picture in Mind

As you progress, keep the big picture in mind. Refer to the storyboard frequently to explain to people how using the structured Value Stream Management process makes extraordinary achievements possible. Major changes to the value stream, combined with small incremental improvements, help create a fast, flexible, customer-driven process with little waste. You can use a "Sunset Report" to document the team's accomplishments and recommend next steps.

What does it mean to have a fast, flexible process? If you are currently scheduling by weeks, releasing by days, and monitoring production by the hour, then work toward scheduling to the day, releasing by the hour, and monitoring production in minutes. If you are measuring cycle times by the minute, measure by seconds. If you measure lead time in weeks, measure in days. If you measure in days, measure in hours!

However, be patient and be realistic. The value stream transformation took four months to complete at the company that provided the data for the Premiere Manufacturing case study. The original time frame was three months, but the team encountered some unexpected obstacles that extended the project's duration. However, the changes made to the target value stream made it possible to operate profitably while still meeting the customer's demand for annual price reductions.

Becoming lean *does* make a difference!

Wrap-Up

Toyota has spent over 50 years perfecting lean manufacturing, and they continue to refine it. For success to occur in your organization, people must continually look for ways to improve the entire value stream. Cultivate this kind of environment daily by recognizing and rewarding people's efforts and treating them with dignity and respect. Be especially open to receiving ideas and suggestions for value stream improvement from operators: they are the ones who create value for your customers and who know the details of the value stream best.

Also, remember that nothing ever goes exactly according to plan. Expect the unexpected, and adjust your plans accordingly.

Good luck on the road to becoming lean!

Glossary of Terms

Autonomation: Automation with a human touch (see *jidoka*). The second of two major pillars of the Toyota Production System. (The first pillar is just-in-time production.)

Benchmarking: A structured approach to identifying a world-class process, then gathering relevant information and applying it within your own organization to improve a similar process.

Buffer inventory (also *buffer stock*): Finished goods available to meet variations in customer demand due to fluctuations in ordering patterns or takt time.

Buffer stock: See *buffer inventory*.

Continuous flow: The ideal state characterized by the ability to replenish a single part that has been "pulled" downstream. In practice, continuous flow is synonymous with just-in-time (JIT) production, which ensures that both internal and external customers receive *only* what is needed, *just when* it is needed, and in the *exact amounts* needed.

Core implementation team: A group of people chartered with planning the details of a Lean manufacturing plan through implementation of the 8-step Value Stream Management process.

Cycle time: The time that elapses from the beginning of a process or operation until its completion.

Demand/customer demand: The quantity of parts required by a customer (see also *takt time*.)

Extended team members: Individuals who provide special skills or expertise to the core implementation team but who are not responsible for implementation.

Flow: The movement of material or information. Manufacturing businesses are successful to the extent that they are able to move material and information with as few disruptions as possible—preferably none.

Heijunka or load leveling: Balancing the amount of work to be done (the load) during a shift with the capacity to complete the work. A heijunka system distributes work in proportions based on demand, factoring in volume and variety.

Heijunka box: A physical device used to level production volume and variety over a specified time period (usually one day). The box is divided into slots that represent pitch increments. The slots are loaded with kanbans that represent customer orders. The order in which kanbans are loaded into the box is determined based on volume and variety.

Jidoka: The second of two Toyota Production System pillars. A method based on the practical use of automation to mistake-proof the detection of defects and free up workers to perform multiple tasks within work cells. In other words, jidoka uses automation in such a way as to promote flow.

Just-in-time production: The first of two Toyota Production System pillars (the other pillar is *jidoka*); a production paradigm which ensures that customers receive *only* what is needed, *just when* it is needed, and in the exact amounts needed. See also *continuous flow.*

Kaizen: Small daily improvements performed by everyone. *Kai* means "take apart" and *zen* means "make good." The point of kaizen implementation is the total elimination of waste.

Kaizen event: A team event dedicated to quick implementation of a Lean manufacturing method in a particular area over a short time period.

Kanban: An inventory control card at the heart of a pull system. The card is a means of communicating upstream precisely what is required (in terms of product specifications and quantity) at the time it is required.

Lean: Shorthand for Lean manufacturing—a manufacturing paradigm based on the fundamental goal of the Toyota Production System: minimizing waste and maximizing flow.

Lean enterprise: An organization that fully understands, communicates, implements, and sustains Lean concepts seamlessly throughout all operational and functional areas.

Leveling: Evenly distributing over a shift or a day the work required to fulfil customer demand. Leveling is achieved either through implementing *paced withdrawal* or *heijunka* (load leveling).

Line balancing: A process in which work elements are evenly distributed within a value stream to meet takt time.

Location indicator: A visual workplace element that shows where an item belongs. Lines, arrows, labels, and signboards are all examples of location indicators.

Muda: See *waste.*

Operator balance chart: A visual display of the work elements, times, and operators at each location in a value stream. The operator balance chart is used to show improvement opportunities by visually displaying each work operation's times in relation to total cycle time and takt time.

Paced withdrawal: A method of leveling that involves moving small batches of material through the value stream over time intervals equal to the pitch.

Pack-out quantity: A small batch equal to the number of units or parts that can be moved throughout the value stream to ensure an efficient flow. Pack-out quantity may or may not be customer driven.

Pitch: The amount of time—based on takt—required for an upstream operation to release a predetermined pack-out quantity of WIP to a downstream operation. Pitch is therefore the product of the takt time and the pack-out quantity.

Product family: A group of parts that share common equipment and processing attributes.

Production kanban: A printed card indicating the number of parts that must be produced to replenish what has been pulled from a supermarket.

Project champion: The person with the authority and responsibility to allocate the organization's resources during the life of the project. The champion should always be completely committed to the project. It is often the champion who initiates the project.

Red tag: A label used in a 5S implementation to identify items that are not needed or that are in the wrong place.

Runner: A worker who ensures that pitch is maintained. The runner covers a designated route within the pitch period, picking up kanban cards, tooling, and components, and delivering them to their appropriate places.

Safety inventory (also *safety stock*): Finished goods available to meet customer demand when internal constraints or inefficiencies disrupt process flow.

Safety stock: see *safety inventory*.

Set in order: The second activity in the 5S system. It involves identifying the best location for each item that remains in the area, relocates items that do not belong in the area, setting height and size limits, and installing temporary location indicators.

Shine: The third activity in the 5S system. It involves cleaning everything thoroughly, using cleaning as a form of inspection, and coming up with ways to prevent dirt, grime, and other contaminants from accumulating.

Signal kanban: A printed card indicating the number of parts that need to be produced at a batch operation to replenish what has been pulled from the supermarket downstream.

Sort: The first activity in the 5S system. It involves sorting through and sorting out items, placing red tags on these items, and moving them to a temporary holding area. The items are disposed of, sold, moved, or given away by a predetermined time.

Standardize (for 5S): The fourth activity in the 5S system. It involves creating the rules for maintaining and controlling the conditions established after implementing the first three S's. Visual controls are used to make these conditions obvious.

Standardized work: An agreed-upon set of work procedures that establishes the best method and sequence for each manufacturing or assembly process. Standardized work is implemented to maximize human and machine efficiency while simultaneously ensuring safe conditions.

Storyboard: A poster-sized framework for holding all the key information for a Lean implementation. It contains the outcomes for each of the 8 steps of Value Stream Management.

Supermarket: A system used to store a set level of finished-goods inventory or WIP and replenish what is "pulled" to fulfill customer orders (internal and external). A supermarket is used when circumstances make it difficult to sustain continuous flow.

Sustain: The fifth activity of the 5S system. It involves ensuring adherence to 5S standards through communication, training, and self-discipline.

Sustain: The fifth activity of the 5S system, where a person or team ensures adherence to 5S standards through communication, training, and self-discipline.

Takt image: The vision of an ideal state in which you have eliminated waste and improved the performance of the value stream to the point that it is possible to achieve one-piece flow based on takt time.

Takt time: The "beat" of customer demand—the time required between completion of successive units of end product. Takt time determines how fast a process needs to run to meet customer demand. Takt time is calculated by dividing the total time available for production by the total customer requirement.

Team charter: A document that includes but that is not limited to the following elements: 1) a clear definition of a team's mission, 2) a statement of team members' roles and responsibilities, 3) a description of the scope of the team's responsibility and authority, 4) project deadlines, 5) a list of metrics and targets, and 6) a list of deliverables (outcomes).

Team leader: The person who facilitates the Value Stream Management process from beginning to end (until a complete kaizen plan is created). The team leader calls and facilitates meetings, ensures that agendas are completed, and manages the allocation and completion of all tasks.

Total cycle time: The total of the cycle times for each individual operation or cell in a value stream. Total product cycle time ideally equals total value-added time.

U-shaped cells: A U-shaped, product-oriented cell layout that allows one or more operators to produce and transfer parts one piece—or one small lot—at a time.

Value stream: A collection of all the steps (both value-added and non-value added) involved in bringing a product or group of products from raw material to finished products accepted by a customer.

Value Stream Management: A sequential, 8-step process used to implement Lean concepts and tools derived from the Toyota Production System. The purpose of Value Stream Management is to minimize the waste that prevents a smooth, continuous flow of product throughout the value stream.

Value stream mapping (or value stream process mapping): The visual representation of the material and information flow of a specific product family; Steps 4 and 6 of the Value Stream Management process.

Withdrawal kanban: A printed card indicating the number of parts to be removed from a supermarket.

Waste (also *muda*): Anything within a value stream that adds cost or time without adding value. The seven most common wastes are 1) waste of overproducing, 2) waste of waiting, 3) waste of transport, 4) waste of processing, 5) waste of inventory, 6) waste of motion, and 7) waste of defects and spoilage.

References

5S and Visual Factory

Galsworth, Gwendolyn D. 1997. *Visual Systems: Harnessing the Power of a Visual Workplace*. New York: American Management Association.

Grief, Michelle. 1991. *The Visual Factory: Building Participation through Shared Information*. Portland, OR: Productivity Press

Hirano, Hiroyuki. 1995. *5 Pillars of the Visual Workplace: The Sourcebook for 5S Implementation*. Portland, OR: Productivity Press.

Peterson, Jim, and Roland Smith. 1998. *5S Pocket Guide*. Portland, OR: Productivity Press.

Productivity Development Team. 1996. *5S for Operators: 5 Pillars of the Visual Workplace*. Portland, OR: Productivity Press.

Productivity Development Team. 2002. *Standardization for the Shopfloor*. Portland, OR: Productivity Press.

Benchmarking

Camp, Robert C. 1989. *Benchmarking: The Search for Industry Best Practices that Lead to Superior Performance*. Portland, OR: Productivity Press.

Watson, Gregory H. 1992. *Benchmarking Workbook: Adopting the Best Practices for Performance Improvement*. Portland, OR: Productivity Press.

Continuous Flow Manufacturing and Cell Design

Hyer, Nancy, and Urban Wemmerlöv. 2002. *Reorganizing the Factory: Competing through Cellular Manufacturing*. Portland, OR: Productivity Press.

Productivity Development Team. 1999. *Cellular Manufacturing: One-Piece Flow for Workteams*. Portland, OR: Productivity Press.

Productivity Development Team. 2002. *Pull Production for the Shopfloor*. Portland, OR: Productivity Press.

Sekine, Keniche. 1992. *One-Piece Flow: Cell Design for Transforming the Production Process*. Portland, OR: Productivity Press.

Continuous Improvement (Kaizen)

Imai, Isaki. 1997. *Gemba Kaizen: A Common-Sense Low-Cost Approach to Management*. New York: McGraw-Hill.

Japan Human Relations Association, eds. 1995. *The Improvement Engine: Creativity and Innovation through Employee Involvement*. Portland, OR: Productivity Press.

———. 1997. *Kaizen Teian 1: Developing Systems for Continuous Improvement Through Employee Suggestions*. Portland, OR: Productivity Press.

———. 1997. *Kaizen Teian 2: Guiding Continuous Improvement through Employee Suggestions*. Portland, OR: Productivity Press.

Laraia, Anthony C., Patricia E. Moody, and Robert W. Hall. 1999. *The Kaizen Blitz: Accelerating Breakthroughs in Productivity and Performance*. New York: John Wiley & Sons.

Productivity Press Development Team. 2002. *Kanban for the Shopfloor*. Portland, OR: Productivity Press.

Rother, Mike, and Rick Harris. 2001. *Creating Continuous Flow*. Brookline, MA: Lean Enterprise Institute.

Suzaki, Kiyoshi. 1987. *The New Manufacturing Challenge: Techniques for Continuous Improvement*. London: Collier Macmillan Publishers

———. 1993. *The New Shop Floor Management*. New York: The Free Press.

Just-in-Time Production

Hirano, Hiroyuki. 1989. *JIT Factory Revolution: A Pictorial Guide to Factory Design of the Future*. Portland, OR: Productivity Press.

———. 1990. *JIT Implementation Manual: The Complete Guide to Just-In-Time Manufacturing*. Portland, OR: Productivity Press.

Productivity Development Team. 1998. *Just-in-Time for Operators*. Portland, OR: Productivity Press.

Kanban

Japan Management Association, eds. 1986. *Kanban and Just-In-Time at Toyota: Management Begins at the Workplace*. Portland, OR: Productivity Press.

Louis, Raymond J. 1997. *Integrating Kanban with MRPII: Automating a Pull System for Enhanced JIT Inventory Management*. Portland, OR: Productivity Press.

Productivity Press Development Team. 2002. *Kaizen for the Shopfloor*. Portland, OR: Productivity Press.

Lean Manufacturing (Lean Production)

Davis, John W. 2001. *Leading the Lean Initiative: Straight Talk on Cultivating Support and Buy-in*. Portland, OR: Productivity Press.

Liker, Jeffrey, ed. 1997. *Becoming Lean: Inside Stories of U.S. Manufacturers*. Portland, OR: Productivity Press.

Womack, James P. and Daniel T. Jones. 1996. *Lean Thinking: Banish Waste and Create Wealth in Your Corporation*. New York: Simon & Schuster.

Womack, James P., Daniel T. Jones, and Daniel Roos. 1990. *The Machine that Changed the World: Based on the Massachusetts Institute of Technology 5-million dollar 5-year Study on the Future of the Automobile.* New York: Rawson Associates

Materials Management and Production Planning

Arnold, J. R. Tony. 1998. *Introduction to Materials Management.* Upper Saddle River, NJ: Prentice Hall

Vollman, Thomas E., William L. Berry, and D. Clay Whybark. 1988. *Manufacturing Planning and Control Systems.* Homewood, IL: Irwin.

Plossl, George. 1983. *Production and Inventory Control: Principles and Techniques.* Atlanta, GA: G. Plossl Educational Services

Mistake-Proofing (Poka-Yoke)

Hinckley, C. Martin. 2001. *Make No Mistake!: An Outcome-Based Approach to Mistake-Proofing.* Portland, OR: Productivity Press.

Nikkan Kogyo Shimbun, eds. 1989. *Poka-Yoke: Improving Product Quality by Preventing Defects.* Portland, OR: Productivity Press.

Productivity Development Team. 1997. *Mistake-Proofing for Operators: The ZQC System.* Portland, OR: Productivity Press.

Shingo, Shigeo. 1986. *Zero Quality Control: Source Inspection and the Poka-Yoke System.* Portland, OR: Productivity Press.

Quality Management

Breyfogle, Forrest W. 1999. *Implementing Six Sigma: Smarter Solutions Using Statistical Methods.* New York: John Wiley & Sons.

Brown, Mark Graham. 2001. *Baldrige Award-Winning Quality, 11th Edition: How to Interpret the Baldrige Criteria for Performance Excellence.* Portland, OR: Productivity Press.

Deming, W. Edwards. 1986. *Out of the Crisis.* Cambridge, MA: Massachusetts Institute of Technology, Center for Advanced Engineering Study

Juran, J. M, ed. 1995. A *History of Managing for Quality: The Evolution, Trends, and Future Directions of Managing for Quality.* Milwaukee, Wis.: ASQC Quality Press.

Quick Changeover

Claunch, Jerry W. 1996. *Setup Time Reduction.* Chicago: Irwin Professional Publishers

Productivity Development Team. 1996. *Quick Changeover for Operators: The SMED System.* Portland, OR: Productivity Press.

Sekine, Keniche, Keisuke Arai, and Bruce Talbot. 1992. *Kaizen for Quick Changeover: Going Beyond SMED.* Portland, OR: Productivity Press.

Shingo, Shigeo. 1985. A *Revolution in Manufacturing: The SMED System.* Portland, OR: Productivity Press.

Teambuilding and Teamwork

Chang, Richard Y. 1999. *Success through Teamwork: A Practical Guide to Interpersonal Team Dynamics*. San Francisco: Jossey-Bass/Pfeiffer.

Kaye, Kenneth. 1994. *Workplace Wars and How to End Them: Turning Workplace Conflicts into Productive Teamwork*. New York: American Management Association.

Larsen, Carl E. and Frank M. Lafasto. 1989. *Teamwork : What Must Go Right, What Can Go Wrong*. Newbury Park, CA: Sage Publications.

Maurer, Rick. 1994. *Feedback Toolkit: 16 Tools for Better Communication in the Workplace*. Portland, OR: Productivity Press.

Michalski, Walter J. 1998. *40 Tools for Cross-Functional Teams: Building Synergy for Breakthrough Creativity*. Portland, OR: Productivity Press.

Scholtes, Peter R., Brian L. Joiner, and Barbara J. Streibel. 1996. *The Team Handbook, Second Edition*. Madison, WI: Joiner.

Thiagarajan, Sivasailam and Glenn M. Parker. 1999. *Teamwork and Teamplay: Games for Building and Training Teams*. San Francisco: Jossey-Bass/Pfeiffer.

Torres, Crecensio, Deborah M. Fairbanks, and Richard L. Roe, ed. 1996. *Teambuilding: The ASTD Trainer's Source Book*. New York: McGraw-Hill.

Total Productive Maintenance

Japan Institute of Plant Maintenance, eds.1997. *Autonomous Maintenance for Operators*. Portland, OR: Productivity Press.

———. 1996. *TPM Encyclopedia*. Atlanta: Japan Institute of Plant Maintenance: Productivity Press.

———. 1997. *Focused Equipment Improvement for TPM Teams*. Portland, OR: Productivity Press.

———. 1997. *TPM for Every Operator*. Portland, OR: Productivity Press.

Koch, Arno. 1999. *OEE Toolkit: Practical Software for Measuring Overall Equipment Effectiveness*. Portland, OR: Productivity Press.

Leflar, James. 2001. *Practical TPM: Successful Equipment Management at Agilent Technologies*. Portland, OR: Productivity Press.

Nachi Fujikoshi Corporation. 1990. *Training for TPM: a Manufacturing Success Story*. Portland, OR: Productivity Press.

Productivity Development Team. 2000. *OEE for Operators: Overall Equipment Effectiveness*. Portland, OR: Productivity Press.

———. 1997. *TPM for Supervisors*. Portland, OR: Productivity Press.

Robinson, Charles J. and Andrew P. Ginder. 1995. *Implementing TPM: The North American Experience*. Portland, OR: Productivity Press.

Sekine, Tenichi and Keisuke Arai. 1998. *TPM for the Lean Factory: Innovative Methods and Worksheets for Equipment Management*. Portland, OR: Productivity Press.

Shirose, Kunio. 1996. *TPM: New Implementation Program in Fabrication and Assembly Industries*. Atlanta: Japan Institute of Plant Maintenance.

———. 1996. *TPM Team Guide*. Portland, OR: Productivity Press.

Tajiri, Masaji and Fumioh Gotoh. 1997. *Autonomous Maintenance in Seven Steps: Implementing TPM on the Shop Floor*. Portland, OR: Productivity Press.

Toyota Production System

Ohno, Taiichi. 1988. *Toyota Production System: Beyond Large-Scale Production*. Portland, OR: Productivity Press.

Monden, Yasuhiro. 1998. *Toyota Production System: An Integrated Approach to Just-In-Time*. Norcross, GA: Engineering and Management Press.

Shingo, Shigeo. 1989. *A Study of the Toyota Production System from an Industrial Engineering* Viewpoint. Portland, OR: Productivity Press.

Value Stream Mapping

Rother, Mike and John Shook. 1999. *Learning to See Version 1.2*. Brookline, Mass.: Lean Enterprise Inst., Inc.

Index